W9-CSP-615

Coliforms and *E. coli*
Problem or Solution?

Coliforms and *E. coli*
Problem or Solution?

Edited by

David Kay
Environment Centre, University of Leeds, UK

Colin Fricker
Thames Water Utilities, Reading, UK

THE ROYAL
SOCIETY OF
CHEMISTRY
Information
Services

Sep/ae
Chem

The Proceedings of the International Conference sponsored by the Water Chemistry Forum of The Royal Society of Chemistry, the Chartered Institution of Water and Environmental Management, Yorkshire Water plc, and Thames Water plc on Coliforms and *E. coli*: Problem or Solution?, held at University of Leeds, 24–27 September 1995.

Special Publication No. 191

ISBN 0-85404-771-9

A catalogue record for this book is available from the British Library.

Published by The Royal Society of Chemistry,
Thomas Graham House, Science Park, Milton Road,
Cambridge CB4 4WF, UK

Printed by Athenaeum Press Ltd, Gateshead, Tyne and Wear, UK.

SD 7/21/97 *twin*

Preface

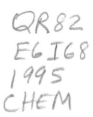

This book derives from a conference organised by the Royal Society of Chemistry and held at Leeds University during September 1995. This was the second in a series of conferences designed to raise awareness amongst professional water scientists of microbiological issues. Sponsoring organisations included the Chartered Institution of Water and Environmental Management, Yorkshire Water, North West Water and Thames Water. The Environment Centre, of Leeds University, provided logistical support. Administration of the conference was handled by Ms Elaine Wellingham.

Over 170 scientists from seventeen countries attended the three day series of presentations and poster sessions. This book contains selected papers and posters organised into five themes; namely,

Detection,
Coliforms and Food,
Natural Waters,
Treated Waters and
Pathogenic *E. coli*.

The book's 29 chapters present review articles and research findings which, together, comprise a state of the art assessment of the significance and utility of this central indicator species and emerging pathogen.

The readership of this book will comprise water scientists and managers, food scientists, academic researchers and students of microbiology.

The Editors and the organising committee are particularly indebted to Ms Elaine Wellingham, the conference organiser, and Mr Trevor Harris who completed desk-top publishing of the manuscript, Mrs Ida Kemp and Joanne Greenwood of Leeds Environment Centre, for their assistance with administration of the event and Mr Duncan McDonald who assisted with final copy editing of this text.

The conference organising committee comprised Dr Clive Thompson, Dr Stuart Clarke, Mr Alan Godfree, Dr Roland Salmon, Dr Gary O'Neill, Dr Bill Keevil, Dr Colin Fricker, Dr Eddie Pike, Dr Michael Waite and Professor David Kay.

David Kay
Colin Fricker

Contents

Detection

Chapter 1
Coliforms: A New Beginning To An Old Problem

Edwin E. Geldreich

Senior Research Microbiologist, National Risk Management Research Laboratory, Cincinnati, Ohio, 45268.

1 INTRODUCTION

Can the coliform indicator concept be "resuscitated" as a better predictor of water quality or are there superior alternatives? For nearly a century, coliform bacteria were thought always to be of intestinal origins and universally assumed to produce acid and gas in lactose fermentation. Coliforms, particularly *E. coli*, are a normal inhabitant of the warm-blooded animal intestinal tract and these bacteria regularly occur in fecal excreta, but there are also many environmental coliforms found in water which have no history of association with fecal material (Knittel *et al.*,1977). While these later investigations have demonstrated the coliform concept to be an over simplification, it still remains a widely accepted descriptor of the microbial quality of water.

Part of the problem is that by definition, coliform bacteria remain an artificial grouping of convenience rather then a precise indication of proven sanitary significance. As a consequence, there has been much uncertainty about coliform significance and there are recurring challenges to their usefulness (Waite,1985). Still, the concept has been a practical tool for providing basic microbial information on water quality. Much of this success has been due to the extensive history of useful application in water supply regulations and legal actions related to management of ambient water quality.

2 CHALLENGE TO SANITARY SIGNIFICANCE

In recent years, this comfortable attitude has been challenged as a result of new information gathered on profiling coliforms in biofilm development, on the discovery of protozoan pathogens in "high quality waters" and on epidemiological investigations in fresh water and marine water recreational areas. These facts complicate the interpretation of coliform significance in many situations. If so, why not abandon the concept and select a more promising candidate?

A reality check suggests any promising new approach has a difficult obstacle course to navigate. Part of the problem lies in the concern that the test must be simple so that all monitoring laboratories can do the examination with minimal equipment requirements and by technicians with only basic skills. The other concern often expressed by water utilities is that a new indicator and associated new standard will be more restrictive rather than more forgiving. These attitudes are driven by a concern that public water supply may not meet the standard imposed by a new indicator system. Such positions may not be tenable if public health protection is at risk and certainly are illogical when compared to the cost of performing examinations for a variety of organic contaminants. The search for a better indicator system or at least a major refinement in the coliform indicator system is in order and long overdue (Geldreich, 1981).

2.1 Alternative Indicator Systems

Other indicator systems have been offered as the ultimate answer, such as fecal streptococci, *Clostridium*, H_2S producers, *Bacteroides*, coliphage and heterotrophic bacteria. Each of these candidates has its champions but each appears to fall short of acceptance for different reasons. Considerations in the evaluations of the candidates include: fecal origins and stability in the aquatic environment; parallel persistence to waterborne agents; occurrence in sufficient densities to measure treatment process efficiency; speciation in the laboratory; rapid testing capability; and monitoring costs. These requirements are very demanding and no known indicator system at this time is immune to some weakness that prevents universal acceptance.

2.2 Recreational Water Criteria

New epidemiological investigations on recreational waters have revealed poor correlations between bather illness and total coliforms (Cabelli *et al.*, 1982; Seyfried *et al.*, 1985; Sanchez *et al.*, 1986; Bernard, 1989; Wyer *et al.*, 1995). These facts support the belief that the total coliform indicator system should be abandoned as worthless. More promising relationships may be found with fecal coliform /*E. coli*, fecal streptococcus/enterococcus and total staphylococcus. Interestingly, indicators also vary in effectiveness as related to fresh vs. marine waters. Much of the evidence points to the need to judge the quality of bathing waters with respect not only to fecal pollution but also to contamination introduced by bathers in close proximity. In fresh bathing waters, fecal coliform/*E. coli* and fecal streptococcus/enterococcus might be used to measure the risk associated with gastrointestinal illness, while total staphylococcus/*Candida albicans* could be used as predictors of illness associated with the mouth, nose and skin areas of bathers (Seyfried *et al.*, 1985). Reports from investigations of marine waters provided conflicting views: one study suggesting a highly significant correlation exists between enterovirus and bacterial indicators (Merrett-Jones *et al.*, 1991) while two other investigator groups found that indicator bacteria (fecal coliforms, fecal streptococci) are poor predictors of enterovirus concentrations (Sanchez *et al.*, 1986; Wyer *et al.*, 1995). Perhaps the more rapid die off of fecal indicator bacteria in marine waters, recirculation of polluted waters near the beach and the protection afforded to viruses in fecal cell debris are important influences on the perception of safe bathing waters.

Coliform bacteria are also a poor measure of the body-contact opportunistic pathogens that might occur in disinfected swimming pools (Geldreich, 1981). As a consequence, there is more interest in the use of enterococci and staphylococci to determine the discrete pockets of epidermal cells, nasal, and saliva contaminants in the surface micro-layer of water in the pool before dissemination. The heterotrophic plate count is an useful indicator of bather load in swimming pools that could have a relationship to bather contamination. These identified indicator systems have a significantly greater resistance to disinfection inactivation than do coliform bacteria, and therefore provide a more realistic characterization of microbial quality.

2.3 Coliform Test Refinements

As noted earlier, coliform bacteria consist of an artificial grouping of organisms believed to be associated with fecal pollution but in reality the target has yet to be hit with any precision. Neither the detection of acid and gas production nor aldehyde development in lactose fermentation at a specified incubation time and temperature includes all coliform strains considered to be of intestinal origin. Since there is evidence that some coliforms do not produce gas as the end product, perhaps the substitution of mannitol for lactose and the recording of acid reaction only would be more inclusive of the organisms of sanitary significance. For example, the search for the pathogen *E. coli* 0157:H7 has introduced a new concern because of the failure of this coliform to grow well in lactose media at 44.5°. This pathogenic *E. coli* will not produce a fluorescence in the 4-methylumbelliferyl-β-D glucuronidase (MUG) assay. However, it will develop a sheen colony on M-Endo medium because of aldehyde production from the metabolism of lactose. By substituting mannitol for lactose, many lactose gas-negative coliforms would be captured in the test along with *Salmonella* and *Yersinia* (Farmer *et al.*,1985). Aldehyde production from mannitol utilization would still produce a sheen colony on M-Endo and an acid reaction from fermentation on M-lauryl sulfate broth. Such a proposed change would not be too radical in light of the recent acceptance of a chemically defined chromogenic coliform test. In this latter approach, the focus has been placed on bacteria possessing the enzyme β-D- galactosidase which is capable of cleaving the chromogenic substrate, thereby releasing the medium chromogen (American Public Health Association,1992). Whichever way a new protocol leads, there will be a regrouping of the inclusion strains of coliform. Hopefully, such a regrouping would be more specific to those bacteria of sanitary significance.

Change in incubation temperature should also be considered in an effort to minimize inclusion of environmental coliforms of no sanitary significance. The traditional use of 35-37°C temperature for coliform cultivation may relate to the human intestinal temperature but may not be the optimum choice for selective growth in a flora of many other organisms. Perhaps new investigation may reveal the advantage of using an elevated temperature in the range of 39-42°C for suppressing interference and providing better selectivity.

2.4 Searching for Rapid Methods

The inability of the laboratory to provide data within a few hours of sample processing has been a major deterrent to greater utilization of microbiology in water plant operations. With the discovery that disinfection by-products may contribute carcinogens and other toxins to water supply, treatment conditions are often modified to minimize these occurrences. In so doing, the margin of safety required to avoid passages of microbial contamination may be compromised and the need for rapid assessment of water quality becomes more urgent than ever. Such information would also be of singular value after contamination occurred either from treatment failure or through line breaks, poor practices in line repair or cross-connections.

Any breakthrough in the development of rapid tests must involve specificity, sensitivity and precision with achievement of a test result within a few hours. Searching for rapid methods in environmental microbiology has revealed a variety of candidate methods that have potential in real time monitoring (Geldreich and Reasoner, 1985). One example that could be cost effective may be seen in the report of a simple membrane filter method incorporating MUG in a modified M-T7 agar medium which provides detection of as few as one fecal coliform per 100 ml in 6 hours (Berg and Fiksdal, 1988). As might be anticipated, rapid tests that can be performed in less than one hour have little specificity and may possibly include non-viable cells. These limitations will eventually be resolved through the discovery of unique media environments that will lead to early growth detection of organisms of interest by instrumentation sensitive to trace concentrations of specific metabolic by-products.

Efforts to shorten incubation time to less than five hours required to detect bacterial indicator systems or individual pathogenic species in a water sample flora will necessitate instrumentation capable of detecting viable organisms directly or after a brief period of amplification in a selective enrichment. These detection systems will become more readily available in the near future; however, achieving routine use will be slow in developing until the systems become cost effective and there is appropriate training for the laboratory staff. The reason for the accelerated interest is the urgent need to know more about treatment barrier effectiveness under restrictions that may be imposed on disinfection concentration to minimize formation of suspect carcinogens.

Rapid testing of recreational water would provide more immediate updates on changing water quality conditions for use in public awareness announcements. Perhaps research should explore the correlation of indicator systems with turbidity concentrations or particulate sizing in local recreational waters, then attempt to use turbidity/particle size as a rapid indicator of changing water quality conditions due to stormwater runoff or sewage discharge migrations. Maybe a threshold level for turbidity/particle size related to rainfall intensity could be used locally to open and close swimming waters. This value could not be used as a national standard because different watershed characteristics, soil types, pollution loadings and climatic conditions affect the threshold level differently at each designated recreational bathing area.

Of the rapid test candidates currently in development, gene probes appear to be the most promising (Richardson *et al.*, 1991). Recent research into microbial genetics has led to techniques that can detect minute quantities of nucleotide sequences unique to a single species of organisms, be they bacteria, fungi, virus or invertebrates. For example, 10 enterotoxigenic *E. coli* per ml of canal water were detected in a grossly polluted canal in Bangkok, Thailand, by using gene probe technology (Moseley *et al.*, 1982).

Use of gene probes and automated microbiological techniques to search for *E. coli* and pathogen passage through treatment barriers, particularly sand filters and disinfectant contact basins, is a major objective. Perhaps multiplex probes to detect bacterial pathogens or enteroviruses will become sensitive enough to detect one organism in a liter of sample within one hour of processing. At the present time, some probe tests can be completed in 30 min to 2 hours but many require 28 to 32 hours for amplification to achieve acceptable sensitivities (Tenover, 1988). Thus it has become apparent that an arsenal of indicator systems may be the trend of the future.

2.5 Winds of Change are Everywhere

The most subtle change over the past 20 years is the general recognition that the coliform test in treated water supply is not so much a measure of sanitary significance as an indication of treatment effectiveness. It is generally agreed that properly operated water treatment systems will inactivate all coliform bacteria, whether they are of sanitary significance or merely environmental coliforms. There is no excuse for coliform bacteria escaping the conventional treatment barrier. However, using coliform bacteria to characterize the effectiveness of water supply treatment does not provide a satisfactory performance standard because these organisms are easily captured in treatment processes (coagulation, settling, filtration) and inactivated by disinfection. As a consequence, attempts to use coliform to gauge adjustments in treatment processes for better removal of protozoan pathogens, viruses and opportunistic pathogens is too coarse a measure. More rigorous tests of treatment effectiveness are being sought. For example, coliphage mimics many of the properties of viruses and is more cost effective to use in the evaluation process. *Clostridium* might be another promising surrogate for testing treatment barriers, although their densities are too low in natural waters to get the desired 6 log removals that might be necessary. Both of these performance indicators will be investigated under the Information Collection Rule now in development by EPA and various water authorities. The use of aerobic sporeformers in the evaluation of treatment processes would appear to be more promising (Rice *et al.*, 1994). Removal of aerobic sporeformers appears to parallel that of particle removals through the treatment train and has the advantage of not breaking apart while passing through various unit processes. For this reason, aerobic sporformers may be a simple surrogate for evaluating the removal efficiencies for environmentally resistant *Giardia* cysts and *Crytosporidium* oocysts. The results to date are very promising; the organisms are common to aquatic waters, have no public health significance and do not multiply in the treatment basin as a result of increased assimilable organic carbon (AOC) after ozonation. Based on this information it becomes apparent that an indicator "cocktail" of coliphage and aerobic sporeformers may be necessary to challenge treatment systems involving physical removal (coagulation/filtration) of cyst-like particles and disinfection for microbial inactivation.

New coliform regulations for drinking water have made a significant impact on public water utilities in several ways. For years, national regulations on drinking water quality for many countries have been based on a standard derived from coliform density measurements gathered over each month to obtain a mean value established at a limit of one coliform per 100 ml. For some American utilities experiencing coliform occurrence, the first reaction to exceeding this limit was to increase the number of samples over the next few days so as to drive the mean value down below the one coliform per 100 ml limit to stay within compliance. This reaction may have avoided a noncompliance issue but did not address the reason why there were coliform bacteria at some location(s) in the distribution system. Regulation revisions by EPA (U.S. Environmental Protection Agency, 1989) changed this approach by defining water quality in terms of coliform frequency of occurrence (presence/ absence) over a 30-day period. A five percent limit is specified and all coliform positive samples have equal weight, so one coliform per 100 ml or several hundred coliforms per 100 ml have the same importance in defining water quality. Unfortunately, the presence/absence approach does not provide information on the magnitude of the contaminating event.

Whenever a distribution water sample is found to contain presumptive evidence of coliforms, regulation requires an immediate resampling at the collection site and several locations above and below the site. The samples are then examined not only for coliform bacteria but also for fecal coliform /E. *coli*. On the repeat finding of a fecal coliform/ E. *coli* at the same site, the water utility must notify the public in that area that the water is unsafe and should be boiled until further notice. If only non-fecal coliform bacteria are found in a sample, public notification is not required, provided these occurrences throughout the distribution system are less than 5 percent for the month. The presumption is that if coliforms are found only occasionally and are not fecal coliforms/E. *coli*, they are probably of environmental origin. Above the 5 percent level, non- fecal coliform occurrences do require public notification but are frequently interpreted to be only a pipe biofilm problem. The danger with this interpretation is that it can hide low level contamination bleeding across the treatment barrier and possible intermittent cross-contamination somewhere in the distribution system.

2.6 New Problems bring New Strategies

With the change in coliform compliance regulations, some of the water utilities that were experiencing biofilm occurrences could no longer circumvent the issue by intensifying the sampling program for the month. At the request of several utilities facing repeated summers of persisting coliform biofilm with no evidence of a waterborne outbreak, an amendment to the Total Coliform Regulation was developed to provide relief from issuing repeated boil-water orders to the community. Avoidance of the noncompliance status required several special conditions to be met including: a weekly check of hospitals for any evidence of increased illness in the community that might be waterborne, a more intense monitoring strategy, profiling of coliform species in water samples and searching for hidden fecal coliform/*E. coli* occurrences that might go undetected in the coliform biofilm event. This latter condition is particularly important since biofilm profiles show a dominance of *Klebsiella, Enterobacter* or *Citrobacter* which could mask the detection of fecal coliform/E.coli organisms in a concurrent fecal contaminating event. This special monitoring strategy reverts to routine monitoring for only total coliform occurrences when the coliform frequency of occurrence eventually falls below the 5 percent level for all samples collected in a month.

Many laboratories have included speciation of coliforms as a consequence of these biofilm events, without realizing that commercial multi-test systems used for speciation were initially developed for use in the clinical laboratory not the aquatic environment. Thus the identification systems may at times fail to provide a suitable substrate for the growth of fastidious species or reaction from slow growing organisms in the aquatic flora.

Heterotrophic bacterial counts in water supply are receiving more recognition as a useful indicator of bacterial colonization within the distribution system. When heterotrophic bacterial populations increase above 1-2 logs or 1,000 organisms per ml., it is time to flush water pipes in the area to reduce sediments, minimize taste and odor problems, and restore a disinfection residual to static water areas of the distribution system. Responding to this early warning of bacterial activity is far more effective than waiting until the pipe network begins to support a biofilm occurrence that eventually includes coliforms. The one major flaw in the use of heterotrophic bacteria counts can be in the selected testing protocol to maximize the recovery of the diverse population of organisms. To maximize detection requires the careful selection of a dilute medium containing a wide range of nutrients, incubation temperature of 28°C and incubation time extended to 5-7 days. One such medium (R-2A) and its 28°C for 5-7 days incubation protocol are being used by numerous water supply utilities. Nutrient-rich agars and one-to-two day incubation at 35°C will not capture many of the organisms in water supply.

Much has been written about the lack of correlation between coliform occurrence and virus detection in water supply. Part of the correlation inconsistency is the lack of evidence that low levels of coliforms may be breaking through the treatment barrier because monitoring data on plant effluents fails to detect any evidence of coliform passage. The heart of the problem appears to lie with three self-imposed limitations: sample volumes standardized at 100ml, infrequent monitoring of the plant effluent, and failure to recover stressed coliforms in disinfected waters. The search for viruses is done by concentrating 100 liters or more of finished water while sample analysis for coliforms appears to be "locked-in" at 100 ml sample portions. If the water supply is thought to be of excellent quality and turbidity is below 0.5 NTU, then why limit the sample size to 100 ml when one liter or more might be more revealing? Many water plants collect only one or two samples from the plant effluent or first customer location each 24- hour period. There is little effort made to use sequential sampling or to correlate sample collections with filter backwashings, turbidity spikes, changes in coagulant dosage, or storm events that also impact on source water quality. Any coliform bacteria not inactivated by disinfectant would be under stress and take time to recover in the distribution system, so samples taken of plant effluent or at the first customer location should be examined on a coliform medium (m-T7 agar) designed to recover stress organisms at these sites (McFeters, 1990).

2.7 New Directions

The continued use of total coliforms should not negate the search for greater acceptance of a diversity of water criteria for use in specific situations. The trend is moving in this direction as we focus on problems with coliform biofilm and its significance, newly identified pathogens (*Giardia*, *Crytosporidium*, *E. coli* 0157:H7), and the growing concern with opportunistic pathogens in both drinking water and recreational waters.

The current trend in criteria diversity is being directed largely to the specialized laboratory with broad expertise and funds to support such investigations. Federal and State Agencies continue to be reluctant to incorporate emerging candidate criteria into regulations and the monitoring programs of the average laboratory. The reasons are varied and worth noting.

Although a wide selection of potential criteria has appeared in the literature, few have been adequately evaluated in the full range of environmental waters to completely establish their indicator significance. Perhaps the EPA Information Collection Rule (ICR) will provide the opportunity to gather a data base on *Clostridium*, coliphage, *Giardia*, and *Crytosporidium* from a variety of source waters and water treatment processing basins that would be difficult to obtain otherwise. Similar information is needed for *Mycobacterium*, fecal streptococci, *Bacteroides*, and staphylococci. In many instances, scientific information is lacking, inadequate, or possibly in conflict, thus precluding a recommendation of specific numerical limits for standard setting or estimation of risk. As a result, when such candidate criteria are used in field situations, interpretation of water quality may be in error and subject to challenge in legal actions. In other instances, attempts have been made to redefine microbial indicator systems without full justification.

The proliferation of methods for a given indicator system can also create problems when evaluating data from different laboratories, particularly when there is no comparative information on the sensitivity, selectivity and reproducibility of different media in an average laboratory. New developments in media must be evaluated in several laboratories in different geographical areas and compared to a standard reference medium. Some of this vital work is being done at the international level (ISO) and in the United States by the American Society for Testing and Materials (ASTM).

Associated closely with general acceptance of the proposed indicator system is the judgement that the procedure can be performed in the average certified laboratory. Such laboratories generally have the capability for performing pour plate, multiple tube, and membrane filter procedures to detect a basic set of microbial indicators and some bacterial pathogens in waters and effluents. Where special skills or additional instrumentation is required to perform the test, there is little chance that the procedure will gain rapid and wide acceptance for use in a network of monitoring laboratories. The major obstacles are the cost of specialized training or certification, associated instrumentation, and an adequate quality assurance program that will provide safeguards for securing reliable data. It is encouraging to note that the interest in *Giardia* and *Crytosporidium* detection by the public water utility laboratory has been so great that all of these obstacles are gradually being overcome for utilities serving over 100,000 people, largely because of the threat of an outbreak.

3 CONCLUSION: ARE COLIFORMS THE PROBLEM OR SOLUTION?

Are total coliform bacteria of public health significance in recreational water and water supply? A growing number of reports would suggest not. In recreational waters, there can be a predominance of environmental total coliform detected that have no correlation to bather illness and therefore are of no public health significance. While indicators of fecal contamination must continue to be used, more attention should be given to a search for opportunistic pathogens through appropriate surrogate indicator systems such as total staphylococcus, *Pseudomonas aeruginosa* and *Candida albicans*. Exposure to opportunistic pathogens in body contact are a major cause of swimmer illnesses and should not be ignored.

In water supply, the answer must be qualified with a "maybe" because of biofilm occurrences in water supply that may harbor the passage of pathogenic contamination. More often, coliform occurrence in treated water supplies is a signal that there may be a breach in the treatment barrier or a loss in the integrity of the distribution system. In either case, a pathway for contaminate passage has occurred that may include some waterborne pathogenic agent. The primary issue is that coliforms are not resistance enough to challenge the effectiveness of treatment barriers to some waterborne pathogens. Will *Clostridium*, coliphage, aerobic sporeformers, or *Mycobacterium* and particulate sizing be the new tools or must we create a genetically engineered "indicator superbug" for evaluating treatment train performance.

Until a better indicator system is established, there are several options worth noting for improving the coliform indicator system in water supply:

1. Establish a tighter limit on coliform releases into the distribution system (measured at the one liter base-line) in an effort to better define effectiveness of surface water treatment and protective soil barriers to ground water. Most systems serving over 10,000 population can probably meet a more restrictive standard, but not so for many of the small water systems using minimal treatment of surface water or no treatment of marginal ground water resources.

2. Profile all coliform occurrences detected in water supply. This trend is not only becoming more common in the water plant laboratory but also of value in the search for *E. coli* and in the verification of a pipe network biofilm. While this supplemental analysis is not required by regulations it has proven very useful in the interpretation of coliform significance.

3. Expand the bacteriological analysis of water through the use of several different criteria for better characterization of source water, process water, finished water and water in the distribution system. This appears to be a growing trend and should be encouraged: fecal coliform analysis in source water, more attention to coliphage and aerobic sporeformer "cocktail" to measure treatment effectiveness in process waters, stressed total coliform detection in finished water plus total coliform detection and heterotrophic bacterial growth in the distribution system to better characterize treatment performance and water quality conditions in the pipe network.

4. Research should investigate the use of mannitol or some other fermentable sugar (as a substitute for lactose) for detecting coliforms of sanitary significance. Furthermore, an effort should be made to optimize recovery of these organisms by investigating incubation temperatures above the traditional 35-37°C. Perhaps a temperature between 39-42°C might better exclude environmental stains of no significance and also suppress many of the interfering heterotrophic bacteria in the water flora.

There is no doubt that microbial characterization of water supply and recreational water needs reform to be more responsive to risks and water quality. We are at a crossroad and it is time to move in the direction towards better defining water quality, through a new beginning to an old problem.

References

American Public Health Association, Standard Methods for the Examination of Water and Wastewater. 18th edition, Amer. Pub. Health Assoc. Washington D.C.,1992.

Berg, J.D. and Fiksdal, L. Rapid Detection of Total and Fecal Coliforms in Water by Enzymatic Hydrolysis of 4- Methylumbelliferone-β-D-Galactoside. Appl.Environ. Microbiol.1988;54:2118-2112

Bernard, A.G. The Bacteriological Quality of Tidal Bathing Waters in Sydney (Australia). Wat.Sci.Tech.1989;21:65- 69.

Cabelli, V.J. Dufour, A.P., McCabe L.J., and Levin, M.A. Swimming-associated gastroenteritis and water quality. Amer.Jour.Epidemiol.1982;115:606-616.

Farmer III, J.J. *et al*. Biochemical Identification of New Species and Biogroups of Enterobacteriaceae Isolated from Clinical Specimens. Jour.Clin.Microbiol.1985;21:46-76.

Geldreich, E.E. Current Status of Microbiological Water Quality Criteria. Amer.Soc.Microbiol.News 1981;47:23-27.

Geldreich, E.E. and Reasoner, D.J. Searching for Rapid Methods in Environmental Bacteriology. In: Rapid Methods and Automation in Microbiology and Immunology. K.O. Habermehl ed. Springer-Verlag, New York. 1985.

Knittel, M.D., Seidler, R.J. and Cabe, L.M. Colonization of the Botanical Environment by Klebsiella Isolates of Pathogenic Origin. Appl.Environ.Microbiol.,1977;34:557-563.

McFeters, G.A. Enumeration, Occurrence, and Significance of Injured Indicator Bacteria in Drinking Water. In: Drinking Water Microbiology, G.A. McFeters ed. pp 478-492. Springer-Verlag, New York,1990.

Merrett-Jones, M. Morris, R., Coope R. and Wheeler, D. The Relationship between Enteric Viruses and Indicators of Sewage Pollution in UK Seawaters. In: Proc. U.K. Symposium on Health-related Water Microbiology. R. Morris, L.M. Alexander, P. Wyn-Jones and J. Sellwood, eds. pp. 158-164. IAWPRC, London,1991.

Moseley, S.L. *et al*. Identification of Enterotoxigenic Escherichia coli by Colony Hybridization Using Three Enterotoxin Gene Probes. Jour.Infect.Dis.,1982;145:863-869.

Rice, E.W., Fox,K.R., Miltner,R.J., Lytle, D.A. and Johnson, C.H. A Microbiological Surrogate for Evaluating Treatment Efficiency. Proc. Water Qual.Tech.Conf., p.2035-2045 Amer.Water Works Assoc. San Francisco, Nov 6-10,1994.

Richardson, K.J, Stewart, M.H. and Wolfe, R.L. Application of Gene Probe Technology for the Water Industry. Jour.Amer.Water Works Assoc.,1991;83:71-81.

Chapter 2
Use Of The Polymerase Chain Reaction To Detect *Escherichia Coli* In Water And Food

E.J.Fricker & C.R.Fricker

Thames Water Utilities, Development Microbiology, Spencer House, Manor Farm Road, Reading, RG2 0JN, U.K.

1 INTRODUCTION

The microbiological examination of drinking water samples to determine the potable nature of the supply, is mostly performed by the membrane filtration (MF) or most probable number (MPN) techniques, although new methods such as those based on defined substrate technology are beginning to be accepted (Fricker *et al*, 1993). Water samples are examined for the presence of indicator bacteria, coliforms and in particular *Escherichia coli*, (Anon, 1994). Detection of *E.coli* is indicative of faecal contamination; other coliforms generally indicate a breakdown in the integrity of the supply system, failure of chlorination or biofilm growth.

Whilst MF and MPN, in particular, have been shown historically to be acceptable techniques, they are nevertheless time consuming and labour intensive, and the final confirmed result may not be available for 72 hours. In addition with the change in definition of coliforms in the U.K., MF especially does not necessarily detect all organisms which fall within the new definition. A number of organisms which are in fact β-galactosidase positive but lactose permease negative will not be identified using standard membrane filtration based methods.

A molecular method, using the polymerase chain reaction (PCR), (Mullis *et al*, 1986) was developed as a rapid confirmatory technique for the detection and identification of *E.coli* isolates, from 44°C incubated membrane filter presumptive colonies. Using PCR as a confirmatory method for *E.coli* from membrane avoids the inherent difficulties associated with the use of PCR for environmental analysis, i.e. sensitivity, inhibition and viability, which may all potentially cause problems with analysis or interpretation of results.

As with current culture methods for water samples, food culture methods suffer from the problem of the time taken to obtain a confirmed result, the labour intensity of the processing and the sensitivity of the methods used. The application of PCR technology to the analysis of food samples would reduce the turnaround time significantly, and decrease labour costs.

This study initially looked at the application of PCR to confirm presumptive food isolates directly from the culture plates. Then in a parallel investigation, methods were developed to try and improve the current culture procedures, by introducing a pre-enrichment step, study the effect on the recovery of *E.coli* from food, and apply PCR to the detection and confirmation of *E.coli* from food samples. A pre-enrichment stage was introduced as a means of improving the sensitivity of the culture method and as a preliminary stage from which potentially multiple PCR applications would be possible. Initial findings from this work are presented here.

2 MATERIALS AND METHODS

The primer sets used were those previously published (Bej *et al*, 1991). They are directed against the *uid A* gene and the *lac Z* gene which code for the production of β-glucuronidase and β-galactosidase, and give products of 154bp and 264bp respectively. The basic protocol was 25 cycles of 1 minute each at 94°C, 60°C and 72°C, with a preliminary 3 minutes at 95°C and a final extension time of 3 minutes at 72°C. The components of the PCR reaction mix were as follows:- dNTP's at 200µM, primers at 0.5µM, MgCl$_2$ at 2.5mM, Tris/HCl at 50mM, KCl at 50mM and Amplitaq at 5U per reaction. A preliminary screening of *E.coli* isolates had shown the primers to detect >98% of the environmental strains tested (Fricker and Fricker, 1994), with this protocol.

2.1 Sample Preparation - Water

The samples were prepared by spiking 10L volumes of tap water with final effluent from sewage works at approximately 1/20 final dilution; chlorinating at 1.5 - 2ppm and removing 1L volumes over a 10 minute time span; neutralising the chlorine in each sample with 200µl of sodium thiosulphate (36g/L).

2.2 Culture Protocol - Water

Volumes (100ml) from the 1L samples were membrane filtered and incubated at 44°C overnight, on filter pads soaked in membrane Lauryl Sulphate Broth (mLSB), according to standard methods (Anon, 1994). Colonies which grew overnight and were yellow to colourless were analysed further. They were resuspended in Diethyl Pyrocarbonate (DEPC) (Sigma Chemicals, UK) water and aliquots (5µl) transferred to standard confirmation media, lactose peptone water (LPW) and tryptone water (TW), and inoculated onto yeast extract agar (YEA), for the cytochrome oxidase test (Anon, 1994). The remainder was then lysed in preparation for amplification by PCR. Some colonies analysed were obtained from membranes following examination of partially treated water. Further identification of colonies was performed, where necessary, using API 20E, (Biomerieux, France).

2.3 PCR Protocol - Water

The isolates for analysis were prepared by lysing the remainder of the suspension used for standard culture analysis in the Thermal Cycler (Perkin-Elmer 480) for 10 minutes at 98°C. An aliquot (25µl) of the lysed suspension was then added to a fresh sterile eppendorf. To this was added PCR reaction mix (150µl) and the whole overlayed with mineral oil (80µl), (Sigma Chemicals). The target DNA was then amplified in a DNA Thermal Cycler (P-E, 480).

Amplified products were visualised on an agarose gel (1.5%) (Pharmacia Biotech, Sweden) containing ethidium bromide at 5µg/ml, and run at 10v/cm for approximately one hour.

The protocol was subsequently adapted to amplification on the Perkin-Elmer 9600 Thermal Cycler. The sample template volume was reduced to 15µl and the PCR reaction mix volume reduced to 90µl. The amplification parameters were also modified to; 95°C for 3 minutes, 94°C for 10 seconds, 60°C for 15 seconds and 72°C for 15 seconds, for 25 cycles, then a final 3 minute extension time at 72°C.

A total of 322 isolates were examined using the P-E 480 protocol and a further 223 isolates examined with the adapted P-E 9600 protocol.

2.4 Sample Preparation - Food

Portions (25g) of turkey and chicken neck skins were stomached with 0.1% peptone water (125ml). The resulting suspensions and body cavity washes were analysed by standard and modified culture methods and PCR.

2.5 Culture Analysis - Food

A total of 70 samples were analysed. Aliquots (0.5ml) of each sample were transferred into single strength MacConkey Broth (MB), (10ml) or single strength Buffered Peptone Water (BPW), (10ml). All samples were incubated overnight at 37°C. Aliquots (100µl) of each overnight culture were inoculated onto MacConkey agar plates and incubated overnight at 44°C. Further aliquots (100µl) of BPW only, were inoculated into fresh MB and incubated overnight at 44°C, then subsequently inoculated onto MacConkey agar. All presumptive *E.coli* colonies isolated on the plates were confirmed by standard methods (Anon, 1994), and PCR. Any which did not give classic confirmation profiles were further identified using Biomerieux API 20E.

2.6 PCR Protocol - Food

Triplicate aliquots (500µl) of each overnight culture were transferred to sterile thin-walled eppendorfs (0.5ml). These were spun at approximately 13,000rpm in a microcentrifuge for 2 minutes. The supernates were removed and one pellet resuspended in DEPC treated water (100µl), for each set of triplicates. The remaining aliquots were retained for back-up analysis. The 100µl suspension was then lysed in the Thermal Cycler at 98°C for 10 minutes. An aliquot (25µl) was then transferred to another thin-walled eppendorf for PCR amplification. PCR mix (150µl) was added and the protocol for P-E 480 followed. All samples were then run on an agarose gel (1.5%) stained with ethidium bromide (0.5µg/ml) to visualise the amplification products.

Further processing of some samples used Chelex as a lysis reagent. Chelex is available as a commercial product, DNA Extraction Reagent (Perkin-Elmer, New Jersey, USA) and it facilitates extraction of DNA from a variety of samples potentially inhibitory to the PCR reaction. Volumes (100µl) were added to one of the triplicate pellets and further processed as for the standard protocol, taking care when removing the 25ml aliquot not to transfer any resin.

A number of samples which showed some discrepancy between results obtained from MB or BPW preliminary culture, and PCR analysis were selected for analysis using an alternative extraction method (Boom *et al*, 1990). This method uses a SiO_2 preparation with the chaotropic agent guanidium thiocyanate to bind the nucleic acids and lyse the cells, and facilitate purification of the DNA, removing potential PCR inhibitors. Pellets (one of the triplicates) were resuspended in aliquots of DEPC water (100µl) as previously and 50µl used for processing with the Boom method. Subsequent aliquots (25µl) were used, after extraction, for amplification using the P-E 480 protocol. In this instance the bands on the gel were present but very faint, so a small aliquot (5µl) was removed from the product vial and added to fresh reaction mix (150µl), then reamplified and run on a fresh gel.

3 RESULTS

3.1 Water Analysis

Initial sample analysis using the P-E 480 protocol showed that eleven samples (3.4%) gave anomalous results on primary analysis when culture and PCR were compared. These are described in detail below.

All 11 samples were identified using the Biomerieux API 20E test. The samples were repeated by culture and PCR analysis, and inoculated onto a glucuronide-containing medium. As a result of this further investigation the following final result was obtained:-

Three samples were "false positive" by PCR; the culture result was indole negative. The API result for one was inconclusive but gave *C.freundii, K.pneumoniae,* and *E.coli* as possible identifications. The second also had a doubtful API profile, including *K.pneumoniae* and *E.coli,* but was glucuronidase positive on the specific medium. The third gave an API profile of *K.pneumoniae pneumoniae.*

Five samples were "false negative " by PCR; the culture result was indole positive. On repeat PCR from the plate inoculum, all confirmed as *E.coli* and they were *E.coli* by API. There was no common factor between these samples although two of the colonies were particularly tiny. They all gave a coliform band and it was the *E.coli* specific band which was not detectable by gel electrophoresis on initial analysis.

Three were "false negative" by standard culture methods; indole negative, they were *E.coli* by API and PCR.

One was "false positive" by culture; the PCR result was negative for *E.coli* and the API result was again doubtful but gave *Citrobacter diversus/amalonaticus* as the most likely. This isolate when tested on glucuronidase specific medium was negative.

All samples (223) analysed using the P-E 9600 protocol showed complete agreement between culture and PCR and no discrepancies were observed.

3.2 Food Analysis

The effect of culturing samples in BPW was compared with the use of MB. For BPW, 23/70 samples yielded *E.coli,* whilst MB recovered *E.coli* from 20/70 samples. These results are summarised in Table 2.1.

Results from samples cultured with a pre-enrichment step in BPW and then MB, followed by inoculating onto MA, were compared with standard culture results. Samples analysed using a pre-enrichment showed a higher positive rate. These results are shown in Table 2.2.

A comparison of PCR analysis results obtained from the two initial culture media, with both DEPC water and Chelex lysis, is shown in Table 2.3. There is no difference in this case between the two media, but analysis using Chelex gave a slightly higher number of positive results.

When the samples were analysed using the Boom method of extraction with double amplification the results were rather unexpected. All samples analysed were positive by PCR (Results not shown) although band intensities varied.

4 DISCUSSION

PCR can be performed in most laboratories after only a small initial outlay and by non-molecular biologists with sufficient training. It is therefore a suitable technique to introduce

Table 2.1 *Detection of E.coli by culture using BPW or MacConkey Broth*

MacConkey	BPW +	-
+	16	4
-	7	43

Table 2.2 *Detection of E.coli with and without pre-enrichment*

Direct	With pre-enrichment +	-
+	20	0
-	8	42

Table 2.3 *Detection of E.coli by PCR using different procedures*

Medium	PCR pre-treatment	Number +ves
MacConkey	None	23
MacConkey	Chelex	25
BPW	None	23
BPW	Chelex	24

for routine analysis. The main problem usually encountered when PCR-based methods are employed, is interpretation of the results. These techniques will detect the DNA from dead cells, so they do not give any indication of viability, and it is possible for the reagents to be inhibited by constituents of some samples, such as humic and fulvic acids and iron. Using the procedure described here for confirmation of presumptive *E.coli* colonies from water sample membranes, there is no subjective interpretation of the results.

The primer sets used during these studies with the protocols described have proved to be dependable for *E.coli* identification. The results obtained with both thermal cyclers were found to be acceptable, and an overall turnaround time, from receipt of sample to confirmed result, of less than 24 hours, is achievable.

The adaptation of the PCR method to the analysis of food samples showed that there is good potential for use of the technique in this area. Whilst there may need to be further improvements in the method, it showed a good correlation with the culture results, whether comparing with the standard method or the pre-enrichment alternative.

The pre-enrichment protocol did in fact improve the recovery by culture. This may be a suitable improvement to the current standard culture method and also an applicable starting point for analysis by PCR, for *E.coli* and other organisms of interest in food analysis. The saving in time to a confirmed result, particularly in the food industry can lead to financial savings and extension to food shelf life.

References

Anon. The microbiology of water, 1994, Part1 - Drinking Water, Report 71, HMSO. 1994.

Bej AK, Mahbubani MH, DiCesare JL, Atlas RM. Polymerase chain reaction-Gene probe detection of microorganisms by using filter-concentrated samples. Appl Environ Microbiol 1991;57:3529-34.

Boom R, Sol CJA, Salimans MMM, Jansen CL, Wertheim-van Dillen PME, van der Noordaa J. Rapid and simple method for purification of nucleic acids. J Clin Microbiol 1990;28:495-503.

Fricker CR, Cowburn J, Goodall T, Walter KS, Fricker EJ. Use of the Colilert system in a large U.K. Water Utility. In: Proceedings of the 1993 Water Quality Technology Conference; Nov 7-11; Miami, Florida: American Water Works Association, 1993:525-7.

Fricker EJ, Fricker CR. Application of the polymerase chain reaction to the identification of *Escherichia coli* and coliforms in water. Lett Appl Microbiol 1994;19:44-6.

Mullis FB, Faloona F, Scharf S, Saiki R, Horn G, Erlich H. Specific enzymatic amplification of DNA *in vitro*: the polymerase chain reaction. Cold Spring Harbor Symp Quant Biol 1986;51:263-73.

Chapter 3
A New Membrane Filtration Medium For The Simultaneous Detection Of Total Coliforms And *Escherichia coli*

M.A. Grant

Hach Co., 100 Dayton Ave., Ames, Iowa, 50010, USA

1 INTRODUCTION

Escherichia coli is used worldwide as the benchmark indicator of fecal pollution (World Health Organization, 1993; American Public Health Association, 1992; Standing Committee of Analysts, 1994). Some other microorganisms possess some of the attributes of *E. coli* as an indicator of fecal influx. These include total coliforms, fecal streptococci and *Clostridium perfringens*. *E. coli* (EC) and total coliforms (TC) have historically been monitored by Multiple Tube Fermentation (MTF) and Membrane Filtration (MF) procedures (International Organization for Standardization, 1990; American Public Health Association, 1992; Standing Committee of Analysts, 1994). Enumeration of EC and TC is useful for reasons stated in the current UK drinking water regulations:

"Regular colony counts provide an indication of overall microbiological quality and each supply zone will have its 'normal' range. Any significant change in this range, suddenly or as a trend over time, and especially of counts at 37°C, will require investigation and, where necessary, remedial action."

(Standing Committee of Analysts, 1994, pg 36).

Membrane filtration procedures have long been used as a simple method to derive enumerative data. Relatively little sample processing is required and these methods are used to analyze water, beverages and some foods. Since current drinking water regulations typically state that 100 ml samples shall contain no TC or EC, it is also possible to use MF in a presence-absence format.

Several media are available for analysis of EC and TC in water and food (World Health Organization, 1984; American Public Health Association, 1992; Standing Committee of Analysts, 1994). One complication with these methods is the necessity to transfer filters from one medium to another and/or from one incubation temperature to another. In some cases incubation of dual filters is required (Standing Committee of Analysts, 1994). Also, some MF methods are designed only for confirmation and not primary isolation (Shadix *et al.*, 1993; USEPA, 1991). Finally, some components of one commonly used MF medium (m-Endo) are believed to be potentially carcinogenic (International Organization for Standardization, 1990).

MF analysis of TC/EC would be facilitated by availability of a medium which would allow simultaneous determination of both populations in a single incubation step of 24 hr or less. The potential utility of such a medium is illustrated by the fact that seven such formulations have been described in the past two years (Sartory and Howard, 1992; Brenner *et al.*, 1993; Chang and Lum, 1994; Hach, 1994; Gelman, 1994; Manafi, 1995; Orenga *et al.*, 1995).

m-ColiBlue24 was developed to allow such simultaneous detection and enumeration. On this medium, TC appear as red colonies and EC as blue colonies. Color remains closely associated with colonies and does not diffuse away from target colonies appreciably. Blue colonies are readily distinguishable even when background TC are TNTC. No UV lamp is required to differentiate TC and EC colonies. To evaluate this new medium, 25 water samples were tested according to a US Environmental Protection Agency protocol (Ulmer, 1992).

2 MATERIALS AND METHODS

2.1 Experimental Protocols

Procedures used in this study were described in the June 30, 1992 version of the US Environmental Protection Agency (USEPA) publication: Requirements for nationwide approval of new or optionally revised methods for total coliforms, fecal coliforms, and/or *E. coli*, in national drinking water monitoring (Ulmer, 1992). Experiments were detailed under the specificity portion of this protocol and are designed to evaluate performance of new media on a presence-absence basis. For each sample, False Positive Error (FPE) and Undetected Target Error (UTE) were determined simultaneously for m-ColiBlue24, m-Endo and mTEC. All isolates picked during these procedures were from plates with 1-5 colonies (Ulmer, 1992, personal communication). Abbreviated outlines of the procedures are as follows. For EC FPE: Target Colony - Endo LES Agar - Nutrient Agar - API20E. For EC UTE: Non-Target Colony - EC Broth - Nutrient Agar - Indole Test - API20E - Return to original medium. For TC FPE: Target Colony - Endo LES Agar - Nutrient Agar - Oxidase Test - LTB - BGLB - API20E (for anaerogenic isolates only). For TC UTE: Non-Target Colony - Nutrient Agar - Gram Stain/Oxidase Test - LTB - BGLB - Return to original medium. For m-ColiBlue24 isolates, EC target colonies were any blue colony and non-target colonies were any non-blue colony. TC target colonies were any red colonies and non-target colonies were colorless colonies. Target and non-target colonies on mTEC, m-Endo and Endo-NA/MUG were as previously described (American Public Health Association, 1992). Calculations were based on the American Society for Testing and Materials, Standard Practice D3870-91 (ASTM, 1991).

2.2 Water Samples

Twenty-five water samples were taken from seven states (California, Colorado, Iowa, Kentucky, New Hampshire, Texas and South Carolina). Standard Methods criteria for holding times were observed (American Public Health Association, 1992). Local samples were returned to the laboratory for immediate analysis and out-of-state samples were shipped with coolant and analysis initiated within 24 hr. Since potable water samples are routinely free of TC and almost invariably free of EC, water samples were used which would provide both target populations. Nineteen were surface water samples, three were non-chlorinated primary effluent from wastewater plants, one was a potable water sample taken during a boil water alert, and two were tap water spiked with wastewater to simulate contamination of drinking water with sewage. For all samples, multiple replicates of several dilutions were filtered onto 0.45 mm GN-Metricel filters (Gelman, Ann Arbor, MI).

2.3 Media

m-ColiBlue24 is supplied in 2-ml ampules (Hach, Loveland, CO). Absorbent pads in Gelman 50 mm MF petri dishes were saturated with m-ColiBlue24 and filters incubated at 35°C for 24±4 hr. m-Endo Broth MF (Difco) was also used to saturate absorbent pads and incubated according to standard methods (American Public Health Association, 1992). mTEC medium (Difco) was used as described (USEPA, 1985) except that the agar was settled out prior to autoclaving and the resulting broth used to saturate absorbent pads. Although mTEC is typically used in the agar-solidified format, this modification was used to provide greater consistency with the m-ColiBlue24 and m-Endo data. Comparisons of recovery using agar solidification and saturated pads were performed with *E. coli* 11775, E. coli 11303, and E. coli 29194. Results showed that recovery was not significantly reduced using the saturated pad format. The technique of Shadix *et al.* (1993), transferring filters from m-Endo to NA/MUG, was also used in nineteen of the samples. FPE and UTE were not determined for this technique, but it was used to compare relative *E. coli* enumeration. m-HPC Agar (Difco) was used to determine relative accuracy of m-ColiBlue24. ISO procedures were as previously described (International Organization for Standardization, 1990). Colonies were examined with a 10X Unitron Stereo Binocular Microscope (Unitron, Plainview, NY). For examination of colony fluorescence during the NA/MUG experiments a 6 watt handheld UV lamp was used.

2.4 Cultures

Coliform cultures used to estimate relative accuracy of m-ColiBlue24 were *E. coli* ATCC 25922, *Enterobacter cloacae* ATCC 23355, *Klebsiella pneumoniae* ATCC 13883 and *Citrobacter freundii* ATCC 8090. Working cultures were maintained in tryptic soy broth (Difco) plus 0.6% yeast extract plus 0.5% agar deeps.

2.5 Relative Accuracy Of m-ColiBlue24

Coliform cultures were incubated 24 hr in TSB+0.6% yeast extract then serially diluted in phosphate-buffered diluent (American Public Health Association, 1992). Quintuplicate 1-ml quantities of appropriate dilutions were filtered and incubated on m-ColiBlue24. A parallel heterotrophic plate count was performed and used to estimate relative accuracy.

2.6 Oxidase Tests

All oxidase results reported in this paper were obtained using standard methods (American Public Health Association, 1992). An alternative oxidase test was also devised during these experiments. This test allowed simultaneous testing of multiple colonies without requiring additional overnight incubation of target colonies struck onto nutrient agar. In this procedure, 0.5-0.75 ml of commercial oxidase reagent (BBL or Difco) were placed in the inverted lid of a standard MF petri dish. A filter, with adhering 24-hr colonies, was lifted from the m-ColiBlue24-saturated pad and placed colony side up onto the reagent in the lid. The reagent soaked up through the filter in approximately 15-20 seconds and oxidase positive colonies either turned blue or exhibited distinct blue coloration adjacent to the colony. This was most easily observed under a dissecting microscope. This version of the oxidase procedure is possible with m-ColiBlue24 since it is not subject to medium acidification (Havelaar *et al.*, 1980; Hunt *et al.*, 1981).

2.7 Statistics

Tests comparing media for Sensitivity, Specificity, False Positive Error, Undetected Target Error and Overall Agreement were carried out using standard statistical procedures for comparison of proportions in independent samples (Snedecor and Cochran, 1989). For example, in comparing the level of Sensitivity in m-ColiBlue24 with that in mTEC, a standard normal deviate was calculated by dividing the difference in observed sensitivities of the two media by the standard error of that difference. The result of the comparison was reported as the probability of a larger standard normal deviate assuming no difference in Sensitivity in the two media.

For comparisons of enumeration, mean colony counts for the media were calculated for each of the 25 samples. Tests of no difference in colony count among media were accomplished using the single criterion of classification analysis of variance. The 25 samples were treated as separate studies. Results were interpreted using treatment means and comparisons among treatment means, along with their standard errors.

3 RESULTS

Table 3.1 compares *E. coli* recovery on m-ColiBlue24 and mTEC. The difference between m-ColiBlue24 and mTEC sensitivity and specificity values were statistically different (p<.05). Similarly, FPE, UTE and Overall Agreement values were statistically different. Table 3.2 indicates total coliform recovery on m-ColiBlue24 as determined by three variations of the experimental protocol. Initial UTE values (Table 3.2A) were acceptable but FPE values were not. Part of the explanation for the high apparent FPE was failure of the protocol to recognize anaerogenic TC. When TC isolates which failed to produce gas in standard LTB/BGLB media were examined by API20E it was found that 25.2% were actually anaerogenic TC. With this correction, the FPE was 26.8%. Further experiments were conducted to see if the oxidase test, as an ancillary procedure, would further reduce FPE. When this data was included, FPE was 17.6%. Table 3.3 indicates total coliform recovery on m-Endo. Interpreting results strictly per protocol gave an FPE value of 43.6%. When correction was made for anaerogenic TC, FPE was 29.0% With API corrections, overall agreement for TC recovery between m-ColiBlue24 and reference methods was 86.2%, and 85.7% with m-Endo. Statistical analysis of the data for sensitivity, specificity, FPE, UTE and overall agreement in Tables 3.2A and 3.3A indicated that only the FPE values were significantly different (p<.05). Statistical analysis of the data in Tables 3.2B and 3.3B indicated none of the five parameters were significantly different.

Results of enumeration data are presented in Table 3.4. Recovery ratios for *E. coli* on m-ColiBlue24 vs. mTEC averaged 3.69 and for m-ColiBlue24 vs Endo-NA/MUG 1.86. Ratios for TC recovery on m-ColiBlue24 vs. m-Endo averaged 3.23.

Comparisons of TC and EC confirmation procedures using US and ISO methods are shown in Table 3.5.

Relative accuracy experiments compared recovery of laboratory-grown coliform cultures on m-ColiBlue24 and the non-selective medium m-HPC. Recovery of *E. coli*, *Enterobacter cloacae* and *Citrobacter freundii* was not significantly different (p<.05) on the two media. Recovery of *K. pneumoniae* ATCC13883 was significantly lower on m-ColiBlue24.

Table 3.1 *Comparison of E. coli Recovery on Two Media*

A. m-ColiBlue24

m-ColiBlue24	Reference positive	negative	total		
positive	234	6	240	Sensitivity	100.0%
				Specificity	97.7%
				FPE [a]	2.5%
				UTE [b]	0%
negative	0	250	250	Overall	
				Agreement	98.8%
	234	256	490		

B. mTEC

mTEC	Reference positive	negative	total		
positive	198	32	230	Sensitivity	97.5%
				Specificity	88.%
				FPE	13.9%
				UTE	2.5%
negative	5	235	240	Overall	
				Agreement	92.1%
	203	267	470		

[a] FPE = False Positive Error
[b] UTE = Undetected Target Error

4 DISCUSSION

Recovery of *E. coli* from a geographically diverse set of water samples was more effective with m-ColiBlue24 than mTEC, as measured by five parameters (Table 1). *E. coli* recovery on m-ColiBlue24 can also be compared to recovery on other recently developed EC/TC media. Brenner *et al.* reported an overall EC specificity value of 95.7% and FPE of 4.3% (Brenner *et al.*, 1993). Orenga *et al.* found 8 of 94 (7.5%) of *E. coli* cultures gave atypical results, i.e. non-pink colonies, on Coli ID (Orenga *et al.*, 1995). Manafi observed that 2.0% of E. coli isolates were GUD negative in LMX-broth (Manafi, 1995).

API20 E codes of E. coli isolates from m-ColiBlue24 and mTEC indicated no obvious difference in the strains recovered. Of the isolates from m-ColiBlue24, 96.5% coded as excellent or very good identifications while 95.9% of mTEC isolates coded as excellent or

Table 3.2 *Total Coliform Recovery on m-ColiBlue24 Using Three Protocol Variations.*

A. Total coliforms identified without correction for anaerogenic isolates.

m-ColiBlue24	Reference positive	negative	total		
positive	120	130	250	Sensitivity	98.4%
				Specificity	65.6%
				FPE	52.0%
negative	2	248	250	UTE	1.6%
				Overall	
	122	378	500	Agreement	73.6%

B. Total coliforms identified using API20E to detect anaerogenic isolates.

m-ColiBlue24	Reference positive	negative	total		
positive	183	67	250	Sensitivity	98.9%
				Specificity	78.7%
				FPE	26.8%
negative	2	248	250	UTE	1.1%
				Overall	
	185	315	500	Agreement	86.2%

C. Total coliforms identified using additional oxidase data.

m-ColiBlue24	Reference positive	negative	total		
positive	206	44	250	Sensitivity	99.0%
				Specificity	84.9%
				FPE	17.6%
negative	2	248	250	UTE	0.96%
				Overall	
	208	292	500	Agreement	90.8%

very good. The number of tryptophanase-negative isolates was also similar: 3.0% from mTEC and 3.3% from m-ColiBlue24.

Visualization of *E. coli* colonies from m-ColiBlue24 was a straightforward procedure,

Table 3.3 *Total Coliform Recovery on m-Endo Using Two Protocol Variations*

A. Total coliforms identified without correction for anaerogenic isolates.

m-Endo	Reference positive	Reference negative	total		
positive	141	109	250	Sensitivity	96.6%
				Specificity	69.2%
				FPE	43.6%
negative	5	245	250	UTE	3.4%
				Overall	
	146	354	500	Agreement	77.2%

B. Total coliforms identified using API20E to detect anaerogenic isolates.

m-Endo	Reference positive	Reference negative	total		
positive	149	61	210	Sensitivity	96.8%
				Specificity	80.1%
				FPE	29.0%
negative	5	245	250	UTE	3.2%
				Overall	
	154	306	460	Agreement	85.7%

with no need to transfer filters or to employ a UV light. Picking of glucuronidase-positive colonies using the NA-MUG procedure (Shadix *et al.*, 1993) was sometimes complicated by diffusion of fluorescent dye away from target colonies and presence of naturally fluorescent non-coliform colonies.

Although this protocol was designed to compare media on a presence-absence basis, preliminary enumeration data indicated m-ColiBlue24 recovered *E. coli* comparably to mTEC and Endo-Nutrient Agar/MUG. Further studies to compare recovery on a quantitative basis are in process.

Recovery of total coliforms on m-ColiBlue24 was similar to recovery on m-Endo. Initial values for UTE were under 2% for m-ColiBlue24 and under 3.5% for m-Endo. These values were acceptable, but initial FPE values were not. TC FPE calculation, strictly per protocol, gave values of 52.0% for m-ColiBlue24 and 43.6% for m-Endo. However, the protocol did not recognize an isolate as a total coliform unless it produced gas in LTB and BGLB. It has been known for several years that true coliforms often fail to produce detectable gas in standard LTB and BGLB (Standridge and Delfino, 1982). The proportion of sheen colonies on Endo which verify by gas production has been reported to vary between 44 and 97% (Evans *et al.*, 1981). Even *E. coli* isolates often fail to produce gas in standard media. Brenner *et al.* reported that 25.8% of their isolates were anaerogenic (Brenner *et al*, 1993). To correct this procedural deficiency, API20E identification was performed on anaerogenic TC isolates. When

Table 3.4 *Ratios of E. coli and Total Coliform Recovery*

Mean recovery ratios (Ave. CFU per 100 ml on m-ColiBlue24/ Ave. CFU per 100 ml on reference medium)	Number of samples	Number of samples in which recovery was significantly different
E. coli Media		
m-ColiBlue24/mTEC = 3.69	23	5
m-ColiBlue24/Endo-NA+MUG = 1.86	19	3
Total coliform media		
m-ColiBlue24/m-Endo = 3.23	25	4

this was done, 25.2% of the putative false positive isolates from m-ColiBlue24 were in fact bone fide TC and 14.6% of the putative false positive isolates from m-Endo were actually TC. With this correction, FPE values were 26.8% for m-ColiBlue24 and 29.0% for m-Endo. These values were not significantly different at $p<.05$.

The total coliform FPE results in this study are in concurrence with previous observations that relatively high FPE values are found among numerous TC-selective media. For example, Jacobs *et al.* used a modification of Clark's formula to analyze TC in a presence-absence format. They reported 10 of 38 isolates were not classical TC, for an FPE of 26.3% (Jacobs *et al.*, 1986). Manafi tested 771 cultures of coliforms and non-coliforms and reported that several species not typically regarded as TC gave positive galactosidase reactions. Specifically, 31 of 34 *Serratia sp.* isolates, 14 of 22 *Hafnia alvei* isolates, 9 of 9 *Aeromonas sp.* isolates and 18 of 47 *Yersinia enterocolitica* isolates (Manafi, 1995).

Similarly, five studies dealing with TC FPE in MF media are illustrative. LeChevallier *et al.* found 12.5% FPE and unidentified isolates with m-T7 (LeChevallier *et al.*, 1983). Brenner *et al.* reported 15.4% of TC target colonies (fluorescent, non-blue) were actually *Serratia sp.* and *H. alvei* and the average coliform verification rate for TC on MI medium was 77.6%(Brenner *et al.*, 1993).

Cenci *et al.* found FPE of 11.7% with m-Endo and 25.1% with MacConkeys plus MUGAL (Cenci *et al.*, 1993). Evans *et al.* noted 14.9% FPE with m-Endo (Evans *et al.*, 1981). Sartory and Howard reported 29.8% FPE with m-LSB, including target colonies which proved to be oxidase positive or both acid and gas negative when transferred to lactose peptone water (Sartory and Howard, 1992). They also cited FPE of 31.2% with m-LGA.

Several studies have also provided information of TC FPE with media used in a MTF format. Lupo *et al.* determined 32-38% of completed MPN values were erroneously inflated due to interference from non-coliforms (Lupo *et al.*, 1977). Edberg *et al.* found FPE of 16% with standard methods and 20% with Colilert (Edberg *et al.*, 1988). Evans *et al.* noted an

Table 3.5 *Analysis of Target Colonies with ISO Media*

A. E. coli

Medium	No. target colonies	No. confirmed by API20E	No. positive in lactose-peptone water	No. positive in tryptone water
m-ColiBlue24	240	234(97.5%)	221(92.1%)	222(92.5%)
mTEC	230	198(86.1%)	200(86.5%)	192(83.5%)

B. Total Coliforms

Medium	No. target colonies	No. confirmed by API20E	No. positive in lactose-peptone water	No. positive in Std. Meth. LTB/BGLB
m-ColiBlue24	250	181(72.4%)	132(52.8%)	120(48%)
m-Endo	250	149(70.9% [a])	180(72.0%)	141(56.4%)

[a] Only 210 m-Endo isolates were analyzed by API20E.

FPE of 5.4% with standard LTB and 12.3% with a modified version (Evans *et al.*, 1981). Palmer *et al.* reported 19% FPE with CL-MW (Palmer *et al.*, 1993). Finally, Covert *et al.* reported that *Serratia sp., H. alvei, Vibrio fluvialis* and *Aeromonas sp.* comprised 25.0% of isolates after 24 hours incubation of Colilert and 40.8% after 28 hours (Covert *et al.*, 1989).

Most of the FPE in these studies, and in the current report, were due to growth of *Serratia sp., Hafnia alvei, Vibrio fluvialis* and *Aeromonas sp.* These bacteria share key physiological similarities with the four traditional coliform genera *Escherichia sp., Enterobacter sp., Klebsiella sp.* and *Citrobacter sp.* It is difficult to insure maximum recovery of these target genera without concomitant growth of some additional Gram negatives. Suppression of FPE at the expense of UTE would be counterproductive. An intrinsic complication is the fact there are 28 species of traditional coliforms, according to Farmer *et al.* (1985), and 17 species according to Brenner (1984). The difficulty of devising growth conditions optimal for all of these, including strains typically stressed by starvation, without permitting growth of some non-coliforms, is evident. In contrast, it is comparatively straightforward to design optimal conditions for selective growth of the single species *E. coli*.

Proposals have been made to expand the current concept of indicator TC to include all *Enterobacteriaceae* (Bonde, 1977; Mossell, 1982). This issue is also addressed in the newest UK methods for drinking water analysis, in which *Hafnia sp., Serratia sp.* and *Yersinia sp.* are recognized as genera which are routinely isolated by methods used for TC and which have similar value as indicators (Standing Committee of Analysts, 1994).

References

American Public Health Association. Standard methods for the examination of water and wastewater. 18th ed. American Public Health Association, American Water Works Association, Water Environment Federation. Washington, DC, 1992.

American Society for Testing and Materials. Standard practice for establishing performance characteristics for colony counting methods in microbiology. D3870-91. In: 1993 Annual book of ASTM standards, Volume 11.02, American Society for Testing and Materials, Philadelphia, 1991.

Brenner DJ. Facultatively anaerobic Gram-negative rods. In: Kreig NR, Holt JG, editors. Bergeys manual of systematic bacteriology. Williams and Wilkins, Baltimore, 1984:408-515.

Brenner KP, Rankin CC, Roybal YR, Stelma GN, Scarpino PV, Dufour AP. New medium for the simultaneous detection of total coliforms and *Escherichia coli* in water. Appl Environ Microbiol 1993;59:3534-44.

Bonde GJ. Bacterial indicators of water pollution. Adv Aquatic Microbiol 1977; 1:273-64.

Cenci G, De Bartolomeo A, Caldini G. Comparison of fluorogenic and conventional membrane filter media for enumerating coliform bacteria. Microbios 1993;76:47-54.

Chang GW, Lum RA. mX, a novel membrane filtration medium for total coliforms and *Escherichia coli* in water. 1994. Q220. Abstracts of the 94th general meeting of the American Society for Microbiology.

Covert TC, Shadix LC, Rice EW, Haines JR, Freyberg RW. Evaluation of the autoanalysis colilert test for detection and enumeration of total coliforms. Appl Environ Microbiol 1989; 55:2443-7.

Edberg SC, Allen MJ, Smith DB, and the National Collaborative Study. National field evaluation of a defined substrate method for the simultaneous enumeration of total coliforms and Escherichia coli from drinking water: comparison with the standard multiple tube fermentation methods. Appl Environ Microbiol 1988;54:1595-1601.

Evans TM, LeChevallier MW, Waarvick CE, Seidler RJ. Coliform species recovered from untreated surface water and drinking water by the membrane filter, standard and modified most-probable-number techniques. Appl Environ Microbiol 1981;41:657-663.

Evans TM, Seidler RJ, LeChevallier MW. Impact of verification media and resuscitation on accuracy of the membrane filter total coliform enumeration technique. Appl Environ Microbial 1981;41:1144-51.

Farmer JJ, Davis BR, Hickman-Brenner FW, McWhorter A, Huntley-Carter GP, Asbury MA, et al. Biochemical identification of new species and biogroups of *Enterobacteriaceae* isolated from clinical specimens. J Clin Microbiol 1985; 21: 46-76.

Federal Register. Drinking water; national primary drinking water regulations; total coliforms (including fecal coliforms and *E. coli*); final rule. Fed Regist 1989;54:27544-68.

Federal Register. National primary drinking water regulations; analytical techniques; coliform bacteria; final rule. Fed Regist 1991; 56:636-43.

Gelman Sciences. MicroSure. PN32639. 1994. Gelman Sciences, Ann Arbor, MI.

Hach Company. New m-ColiBlue24 broth. Literature No. 4331. 1994. Hach Company, Loveland, CO.

Havelaar AH, Hoogendorp CJ, Wesdorp AJ, Scheffers WA. False-negative oxidase reaction as a result of medium acidification. Ant van Leeuwen 1980; 46:301-12.

Hunt LK, Overman TL, Otero RB. Role of pH in oxidase variability of *Aeromonas hydrophila*. J Clin Microbiol 1981;13:1054-9.

International Organization for Standardization. Water quality - detection and enumeration of coliform organisms, thermotolerant coliform organisms and presumptive *Escherichia coli*. Part 1 and 2. ISO 9308-1:1990(E), ISO 9308-2:1990(E). 1990. International Organization for Standardization, Geneva.

Jacobs NJ, Zeigler WL, Reed FC, Stukel TA, Rice, EW. Comparison of membrane filter, multiple-fermentation tube, and presence-absence techniques for detecting total coliforms in small community water systems. Appl Environ Microbiol 1986; 51:1007-12.

LeChevallier MW, Cameron SC, McFeters GA. New medium for improved recovery of coliform bacteria from drinking water. Appl Environ Microbiol 1983; 45:484-92.

Lupo L, Strickland E, Dufour A, Cabelli V. The effect of oxidase positive bacteria on total coliform density estimates. Health Lab Sci 1977;14:117-21.

Manafi M. New medium for the simultaneous detection of total coliforms and *Escherichia coli* in water. 1995. P43. Abstracts of the 95th general meeting of the American Society for Microbiology.

Mossell DAA. Microbiology of foods: the ecological essentials of assurance and assessment of safety and quality. Utrecht, University of Utrecht, 1982.

Orenga S, Barbaux L, Gayrol JP, Lhopital MA, Villeval F. Evaluation of Coli ID, a new chromagenic medium for enumeration and detection of *Escherichia coli* and coliforms. 1995. P42. Abstracts of the 95th general meeting of the American Society for Microbiology.

Palmer CJ, Tsai YL, Lang AL, Sangermano LR. Evaluation of Colilert-Marine Water for detection of total coliforms and *Escherichia coli* in the marine environment. Appl Environ Microbiol 1993; 59: 786-90.

Sartory DP, Howard L. A medium detecting β-glucuronidase for the simultaneous membrane filtration enumeration of *Escherichia coli* and coliforms from drinking water. Lett Appl Microbiol 1992; 15:273-6.

Shadix LC, Dunnigan ME, Rice EW. Detection of Escherichia coli by the nutrient agar plus 4-methylumbelliferyl β-D-glucuronide (MUG) membrane filter method. Can J Microbiol 1993; 39: 1066-70.

Snedecor GW, Cochran WG. Statistical Methods. 8th ed. 1989. Iowa State University Press, Ames, IA.

Standing Committee of Analysts. The microbiology of water 1994. Part 1-drinking water. Report on public health and medical subjects No 71. Methods for the examination of waters and associated materials. 6th ed. HMSO, London.

Standridge, JH, Delfino, JJ. Underestimation of total-coliform counts by the membrane filter verification procedure. Appl Environ Microbiol 1982;44:1001-3.

Ulmer NJ. Requirements for nationwide approval of new or optionally revised methods for total coliforms, fecal coliforms, and/or *E. coli*, in national drinking water monitoring. 1992. Revision 1.2. US Environmental Protection Agency, Environmental Monitoring Systems Laboratory, Cincinnati, Ohio.

Ulmer NJ. 1992. Personal communication.

US Environmental Protection Agency. Test methods for *Escherichia coli* and enterococci by the membrane filter procedure. 1985. Publication EPA-600/4-85-076. Environmental Monitoring and Support Laboratory, US Environmental Protection Agency, Cincinnati, Ohio.

US Environmental Protection Agency. Test methods for *Escherichia coli* in drinking water. 1991. Publication EPA/600/4-91/016. Environmental Monitoring Systems Laboratory, US Environmental Protection Agency, Cincinnati, Ohio.

World Health Organization. Guidelines for drinking-water quality. Volume 1. Recommendations. 2nd ed. 1993. World Health Organization, Geneva.
World Health Organization. Guidelines for drinking-water quality. Volume 2. Health criteria and other supporting information. 1984. World Health Organization, Geneva.

Chapter 4
Evaluation Of Proprietary Culture Media For The Simultaneous Detection Of Coliforms And *E.Coli* Using The RABIT System

Andrew Pridmore & Peter Silley

Don Whitley Scientific Limited, 14 Otley Road, Shipley, West Yorkshire, BD17 7SE, UK

1 INTRODUCTION

Monitoring of drinking water for the presence of specific pathogenic bacteria is impracticable and indeed unnecessary for routine quality control purposes. Any pathogenic microorganisms present in water are usually greatly outnumbered by normal commensal flora of the human or animal gastro-intestinal tract. A more logical approach is the detection of organisms normally present in the faeces of humans and other warm-blooded animals, as indicators both of faecal pollution and of the efficacy of water treatment.

In relation to public health the principal tests applied to water are the viable plate count and those for coliforms, faecal streptococci and sulphite-reducing clostridia. Coliforms and *Escherichia coli* are the most widely used indicators of faecal pollution, as this group offers a relatively simple and cost effective means of assessing contamination. Even so, this important indicator system has recently been subjected to intensive re-evaluation, with the objective of improving both the speed and specificity of test methods. Currently, the most widely accepted technique for the detection of coliforms and thermotolerant coliforms (presumptive *E.coli*) is membrane filtration. Although convenient for testing large numbers of water samples, the method requires separate membranes for total and thermotolerant coliforms, both of which are incubated for 24h to obtain presumptive counts. A further 24h is then required for confirmatory tests to be performed.

Recently, culture media have become available which contain the components required for the performance of confirmatory tests after the initial 24h incubation. Merck "Fluorocult" media offer this facility by the incorporation of 4-methylumbelliferyl-ß-D-glucuronide (MUG) and tryptophan into conventional formulations. The MUG substrate is utilized almost exclusively by *E.coli* to form methylumbelliferone, which fluoresces under long-wave UV light. An indication of the presence of *E.coli* can therefore be obtained after 24h. Tryptophan is metabolised by *E.coli* to produce indole, which can also be detected after 24h as a further confirmation of *E.coli* presence. The two "Fluorocult" formulations applicable to water testing include Fluorocult BRILA Broth, based on brilliant green bile broth, and Fluorocult Lauryl Sulphate Broth.

Recent changes in guidelines for the identification of coliforms and *E.coli* in water (eg Anon, 1994) have led to the introduction of Colilert media (IDEXX laboratories). These novel formulations contain o-nitrophenyl-ß-D-galactopyranoside (ONPG) in addition to MUG. ONPG is cleaved by the ß-galactosidase enzyme possessed exclusively by coliforms, yielding the bright yellow product o-nitrophenol. An appropriate colour change in the incubated medium thus confirms the presence of coliforms in the sample. Specific confirmation of *E.coli* presence is also possible due to the utilization of MUG by this organism as described previously.

The Fluorocult media are liquid formulations, and Colilert is a dry formulation for direct addition to water samples. Both types of medium therefore offer a presence/absence test but cannot normally provide quantitative results. The present study investigated the possibility of enumerating water-borne bacteria by using these media in conjunction with the Rapid Automated Bacterial Impedance Technique (RABIT). The RABIT system detects bacteria on the basis of the electrical changes which they produce in the culture medium during growth, or alternatively by detection of metabolic carbon dioxide. In both cases, detection by impedance can be correlated with numerical results obtained using conventional methods.

This study therefore compared Fluorocult and Colilert media, used in RABIT, with conventional membrane filtration methods for the detection of total coliforms and *E.coli* in water.

2 MATERIALS AND METHODS

2.1 Culture media and Equipment

Merck Fluorocult BRILA Broth and Fluorocult Lauryl Sulphate Broth (Catalogue numbers 12587 and 12588 respectively) were obtained from BDH Limited, Poole, UK. Colilert "Regular" and Colilert 18 media (Catalogue numbers WP020 and WP200 respectively) were obtained from IDEXX Laboratories Limited, Chalfont St Peter, UK. Cellulose nitrate filter membranes of diameter 47 mm and pore size 0.45 mm (Catalogue number 7141114) were obtained from Whatman International Limited, Maidstone, UK.

2.2 Bacterial Strains

Methods for the detection of coliforms were evaluated using *Klebsiella aerogenes* DWC 0056, obtained from the Don Whitley Scientific culture collection. Detection of *Escherichia coli* was evaluated using *E.coli* ATCC 25922 and *E.coli* NCTC 10418, obtained from the respective type culture collections. The specificity of detection methods was assessed by the inclusion of *Pseudomonas aeruginosa* ATCC 10145 as a representative water-borne organisms.

2.3 Procedure

All bacterial strains were maintained in pure culture on Plate Count Agar (LabM; LAB149) at 37°C. Subcultures were made into Tryptone Soy Broth (LabM; LAB4) and incubated at 37°C overnight to provide inocula for each experiment.

Two methodologies were used with the RABIT system; in the direct impedance technique, culture media and bacterial inoculum were simply placed in the impedance cell in direct contact with the electrodes at the base of the cell. Bacterial growth was then detected on the basis of positive conductance changes occurring in the culture media during incubation. In the indirect impedance technique, the cell electrodes were connected by an agar bridge containing potassium hydroxide. The culture medium and inoculum were placed in a separate insert which was positioned above the electrodes. The indirect cell was then tightly stoppered and bacterial growth was detected on the basis of negative conductance changes occurring in the agar bridge when metabolic carbon dioxide was absorbed during incubation following microbial growth. The results generated from both methods were expressed as Time-to-Detection (TTD) which represents the point at which the rate of change of conductance across the test cell exceeds the detection threshold pre-set by the operator. It is this parameter which can be correlated with conventional enumeration methods: when the test system is optimized, TTD shows a strong negative correlation with inoculum level and can therefore be used to calibrate the RABIT system, permitting automatic quantitative results to be obtained from subsequent tests.

Initial experiments evaluated Merck "Fluorocult" media for the detection of coliforms and *E.coli* using direct and indirect impedance technology. Fluorocult Lauryl Sulphate Broth (FLSB) was found to be unsuitable for use in direct impedance work due to its high electrical conductivity, and was therefore evaluated using the indirect technique only. Dense overnight broth cultures of *E.coli* ATCC 25922 and *K.aerogenes* DWC 0056 were prepared as described above, then each was serially diluted in sterile Maximum Recovery Diluent (MRD; Don Whitley Scientific Limited) to a level of 10^{-7}. The actual inoculum levels provided by the diluted cultures were determined by surface inoculation of Plate Count Agar with the 10^{-4} dilutions of each broth using a Model D Spiral Plater (Don Whitley Scientific Ltd). Spiral plates were incubated at 37°C overnight then enumerated in accordance with the manufacturer's instructions. BRILA broth and FLSB were prepared following the label instructions, then dispensed in 5 ml aliquots into sterile RABIT impedance cells (BRILA broth) or sterile glass inserts for the indirect RABIT method (BRILA and FLSB). Duplicate cells of each medium were inoculated with 0.1 ml aliquots of 10^{-1}, 10^{-3}, 10^{-5} and 10^{-7} dilutions of each culture to provide estimated inoculum levels in the range 10 to 10^{7} colony-forming units (cfu). Uninoculated blanks were also included for each medium.

All cells were placed in RABIT incubator modules set at 37°C and incubated using a pre-set RABIT Test Code containing the following parameters:

Test Duration	:	24h
Time Resolution	:	6 min
Temperature	:	37°C
Detection criterion	:	+10mS (direct method)
		-10mS (indirect method)

On completion of the RABIT tests all cells were removed from the modules and illuminated with a long-wave (366nm) ultra violet lamp. The presence or absence of fluorescence in the incubated media was recorded. If necessary the exercise was repeated after transferring the contents of the incubated cells to clear polystyrene tubes.

An indole test was also performed on the incubated Fluorocult media in accordance with the manufacturer's instructions: a thin layer of Kovac's reagent was pipetted on to the surface of the medium within the RABIT cell and examined for 2 minutes. A colour change to red was recorded as a positive indole test.

Colilert media allow the detection of coliforms on the basis of ß-galactosidase activity, in accordance with recent guidelines, and were therefore evaluated more extensively in the RABIT system. Two formulations were available: Colilert "Regular" normally requires 24h incubation whereas Colilert 18 has a different substrate profile and is said to provide reliable results after 18h. Dense overnight broth cultures of *E.coli* ATTC 25922, *K.aerogenes* DWC 0056 and *Pseudomonas aeruginosa* ATCC 10145 were prepared as described previously, then each was serially diluted in sterile MRD to a level of 10^{-8}. The actual inoculum levels provided were determined by enumerating the 10^{-4} dilution as described above. Colilert and

Colilert 18 were prepared by dissolving the contents of the single-dose snap-packs in 100 ml volumes of sterile deionized water, and were dispensed in 5 ml aliquots into sterile glass inserts for the indirect RABIT method. Duplicate cells of each formulation were inoculated with 0.1 ml aliquots of 10^{-6} - 10^{-8} dilutions to provide estimated inoculum levels in the range 1-100 cfu, considered to be representative of the levels to be detected in 100 ml samples from potable water supplies. Uninoculated blanks were also included.

All cells were placed in RABIT incubator modules set at 35°C, the recommended temperature for Colilert media, and incubated using the parameters described previously for the indirect RABIT method. On completion of the RABIT tests all cells were removed from the modules and the colour of the incubated media recorded. Cells were then exposed in LW-UV light and any fluorescence recorded.

Colilert 18 was subjected to a more detailed evaluation in RABIT by performing a comparison with the standard membrane filtration technique for enumeration of coliforms in water samples. Serial dilutions of *E.coli*, *K.aerogenes* and *P.aeruginosa* broth cultures were prepared and used to inoculate 300 ml volumes of autoclaved tap water, to provide samples containing 10-1000 cfu per 100 ml. A proportion of samples contained mixtures of organisms in order to evaluate recovery of low coliform numbers in the presence of higher levels of *P.aeruginosa*, and low *E.coli* numbers in the presence of higher total coliform levels. The combinations tested are shown in Table 5. Each water sample was divided into 3 x 100 ml portions which were filtered through sterile cellulose nitrate membranes (pore size 0.45 mm) using a stainless steel three-funnel filtration manifold and vacuum pump. One membrane from each sample was transferred to a sterile glass insert containing 5 ml of Colilert 18, and tested using the indirect RABIT method as described above. The two remaining membranes were transferred to absorbent pads saturated with membrane lauryl sulphate broth in the base of sterile Petri dishes. Both membranes were incubated at 30±1°C for 4h, then one was transferred to 37±1°C for 18h, the other to 44±0.5°C for the same period, in order to recover total coliforms and thermotolerant coliforms respectively. After incubation, all yellow colonies developing on the membranes were enumerated and the results expressed as cfu per 100 ml of water.

The final experiment compared the RABIT/Colilert 18 method with membrane filtration for the recovery of coliforms from water after exposure to chlorine. Serial dilutions of *E.coli* and *K.aerogenes* broth cultures were prepared and used to inoculate 300 ml volumes of water to provide samples containing 0-100 cfu per 100 ml. Selected samples were prepared in triplicate: of these, one was tested without further treatment, while to the other two were added dilutions of sodium hypochlorite to provide available chlorine concentrations of 1 and 5 ppm respectively. Chlorinated samples were mixed for 15 minutes, after which 0.3 ml of sterile 3.9% sodium thiosulphate solution was added to each 300 ml in order to neutralize residual chlorine. All samples were then filtered in 3 x 100 ml aliquots as described previously. One of the three membranes was incubated in RABIT using Colilert 18, the other two were enumerated after incubation on MLSB at 37°C and 44°C as before.

3 RESULTS

The data generated from Fluorocult BRILA broth in the RABIT system are shown in Table 4.1 (direct method) and Table 4.2 (indirect method). In both tables results are expressed as Time-to-Detection (TTD) and as Total Change, the latter representing the total conductance change produced in the test cell after 24h incubation. In both methods a strong impedance signal was obtained from the growth of *E.coli* and *K.aerogenes*, as demonstrated by the difference in Total Change between uninoculated blanks and positive cells. The larger changes produced in the indirect method are indicative of the large volumes of CO_2 produced by these organisms. Both methods also produced reproducible TTD results from each inoculum, the figures for each culture increasing with dilution factor. This allowed a strong correlation to be made between TTD and inoculum level using the RABIT calibration software, the calibration graph produced using the direct method is shown in Figure 4.1.

In cells inoculated with *E.coli,* strong fluorescence was visible under LW-UV light and a positive indole test could be read within 30 seconds of adding Kovac's reagent. Fluorescence was best observed after transferring the incubated medium to a clear container, but the indole test could be read in the direct or indirect RABIT cell without difficulty.

The results obtained from Fluorocult Lauryl Sulphate Broth in the indirect RABIT method are presented in Table 4.3. A strong impedance signal was also produced in this medium from *E.coli* and *K.aerogenes*, and reproducible TTD figures allowed a strong correlation to be made with inoculum level (Figure 4.2). Fluorescence at 366 nm and a positive indole test were obtained from cells inoculated with *E.coli.*

Table 4.4 shows the comparison between Colilert "Regular" and Colilert 18 for the detection of pure culture in the indirect RABIT technique. Large conductance changes were produced by coliform growth in both media, and reproducibility of Time-to-Detection for duplicate tests was also good. Colilert 18 allowed detection of a given inoculum 3 to 4 hours sooner than the original formulation. *Pseudomonas aeruginosa* inocula of up to 1500 cfu

Table 4.1 *Detection of Escherichia coli ATCC 25922 and Klebsiella aerogenes DWC 0056 in the direct RABIT method using Fluorocult BRILA broth*

Organism and dilution	Inoculum (cfu/0.1ml)	Time to detection (HH:MM)	Total Change (μS)	Fluorescence ($\pm$)	Indole ($\pm$)
E.coli 10^{-1}	1.5×10^7	2:24 3:12	1204 1391	+ +	+ +
E.coli 10^{-3}	1.5×10^5	4:48 4:30	1319 1403	+ +	+ +
E.coli 10^{-5}	1.5×10^3	6:36 6:36	1299 1404	+ +	+ +
E.coli 10^{-7}	1.5×10^1	8:30 9:00	1831 1431	+ +	+ +
K.aerogenes 10^{-1}	1.9×10^7	3:00 3:06	1456 1633	- -	- -
K.aerogenes 10^{-3}	1.9×10^5	5:12 5:18	1530 1469	- -	- -
K.aerogenes 10^{-5}	1.9×10^3	7:36 7:36	1627 1332	- -	- -
K.aerogenes 10^{-7}	1.9×10^1	9:24 9:54	1441 1446	- -	
BLANK	0	>24 >24	219 177	- -	- -

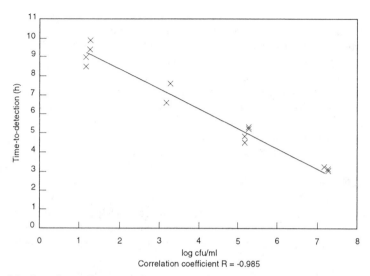

Figure 4.1 *Correlation between platecount and RABIT (BRILA Broth, direct method) for enumeration of coliforms*

Table 4.2 *Detection of Escherichia coli ATCC 25922 and Klebsiella aerogenes DWC 0056 in the indirect RABIT method using Fluorocult BRILA broth*

Organism and dilution	Inoculum (cfu/0.1ml)	Time to detection (HH:MM)	Total Change (μS)	Fluorescence (±)	Indole (±)
E.coli 10^{-1}	1.5×10^7	1:30	6286	+	+
		2:30	5371	+	+
E.coli 10^{-3}	1.5×10^5	4:18	5946	+	+
		4:18	5283	+	+
E.coli 10^{-5}	1.5×10^3	6:06	5775	+	+
		6:18	5894	+	+
E.coli 10^{-7}	1.5×10^1	8:12	6073	+	+
		8:18	6130	+	+
K.aerogenes 10^{-1}	1.9×10^7	1:24	6115	-	-
		1:24	5913	-	-
K.aerogenes 10^{-3}	1.9×10^5	3:18	5883	-	-
		3:18	6131	-	-
K.aerogenes 10^{-5}	1.9×10^3	5:36	5868	-	-
		5:48	5820	-	-
K.aerogenes 10^{-7}	1.9×10^1	7:18	3549	-	-
		7:30	4771	-	-
BLANK	0	>24	1086	-	-
		>24	1714	-	-

were not detected in either media, demonstrating selectivity for the target organisms. Coliform detection was further confirmed by a change in the incubated medium from colourless to yellow. All cells in which growth of *E.coli* was detected (at an inoculum level of 2 cfu in one case) also exhibited fluorescence when exposed to long wave UV light.

Results obtained from the comparison of Colilert 18 in the RABIT system with conventional membrane filtration for testing artificially contaminated water samples are shown in Table 5. These data show that RABIT has a similar sensitivity to the conventional method for the detection of low coliform numbers. The inclusion of *P.aeruginosa* in some of the samples did not interfere with detection or with the indicator system of the Colilert 18: all cells in which coliform growth was detected showed a yellow colouration at the end of the incubation period and all positive *E.coli* cells exhibited fluorescence.

The chlorination experiments generated few results, due partly to the high dilution of some of the inocula and also the apparent effect of chlorine on the low numbers of organisms. However, one of the samples yielding 15 thermotolerant coliforms per 100 ml after exposure to 1 ppm of chlorine was detected in RABIT and exhibited both a colour change and fluorescence. The original inoculum of *E.coli* added to this sample was calculated as <1 cfu/ 100 ml.

The RABIT TTD results obtained from pure cultures and water samples tested using Colilert 18 (Tables 4.4 and 4.5) were combined and correlated with plate counts and membrane filter counts using the RABIT calibration software - this calibration graph is shown in Figure 4.3, which reveals a significant ($p = 0.05$) correlation between TTD and bacterial numbers.

Table 4.3 *Detection of Escherichia coli ATCC 25922 and Klebsiella aerogenes DWC 0056 in the indirect RABIT method, using Fluorocult Lauryl Sulphate Broth*

Organism and dilution	Inoculum (cfu/0.1ml)	Time to detection (HH:MM)	Total Change (μS)	Fluorescence ($\pm$)	Indole ($\pm$)
E.coli 10^{-1}	1.5×10^7	1:54	3950	+	+
		1:54	3625	+	+
E.coli 10^{-3}	1.5×10^5	4:12	3568	+	+
		4:12	2481	+	+
E.coli 10^{-5}	1.5×10^3	6:42	3701	+	+
		6:30	3257	+	+
E.coli 10^{-7}	1.5×10^1	8:36	3854	+	+
		8:36	3765	+	+
K.aerogenes 10^{-1}	1.9×10^7	1:24	3812	-	-
		1:24	3434	-	-
K.aerogenes 10^{-3}	1.9×10^5	3:24	3237	-	-
		3:24	3728	-	-
K.aerogenes 10^{-5}	1.9×10^3	5:48	3590	-	-
		5:42	2828	-	-
K.aerogenes 10^{-7}	1.9×10^1	8:18	3320	-	-
		8:30	3450	-	-
BLANK	0	>24	1248	-	-
		>24	1449	-	-

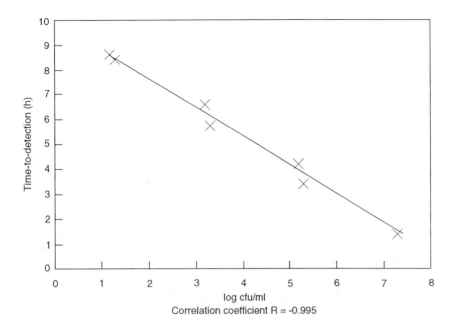

Correlation coefficient R = -0.995

Figure 4.2 *Correlation between platecount and RABIT (FLSB, indirect method) for enumeration of coliforms*

Table 4.4 *Selective detection of coliforms and Escherichia coli using Colilert media in the indirect impedance technique*

Culture	Dilution	Inoculum (cfu per RABIT cell)	Colilert Regular				Colilert 18			
			TTD (HH:MM)	Total Change (μS)	Yellow Colour	Fluoresc-ence	TTD (HH:MM)	Total Change (μS)	Yellow Colour	Fluoresc-ence
E.coli ATCC 25922	10^{-6}	200	12:18 12:30	3098 3406	+ +	+ +	9:30 9:36	2728 2638	+ +	+ +
	10^{-7}	20	14:12 14:54	2646 2983	+ +	+ +	10:00 11:24	2353 1964	+ +	+ +
	10^{-8}	2	>24 16:30	493 1909	- +	- +	>24 >24	986 693	- -	- -
K.aerogenes DWC 0056	10^{-6}	100	13:12 13:36	2087 2182	+ +	- -	9:24 9:54	2461 2247	+ +	- -
	10^{-7}	10	15:48 14:54	1873 1889	+ +	- -	11:18 11:48	2118 1960	+ +	- -
	10^{-8}	1	17:06 >24	1598 725	+ +	- -	>24 16:18	771 479	+ +	- -
P.aeruginosa ATCC 10145	10^{-6}	1500	>24 >24	668 823	- -	- -	>24 >24	520 799	- -	- -
	10^{-7}	150	>24 >24	599 837	- -	- -	>24 >24	566 549	- -	- -
	10^{-8}	15	>24 >24	655 650	- -	- -	>24 >24	514 747	- -	- -

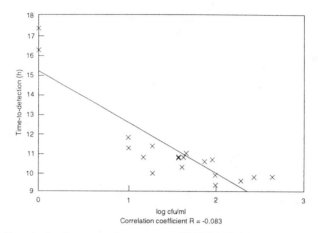
Correlation coefficient R = -0.083

Figure 4.3 *Correlation between platecount and RABIT (Colilert 18, indirect method) for enumeration of coliforms*

Table 4.5 *Detection of coliforms and E.coli in normal and hyper-chlorinated tap water using Colilert 18 in the indirect impedance technique*

Cultures added	Inoculum (cfu/100 ml)	Added Chlorine (ppm)	Membrane filtration results (cfu/100 ml)		RABIT TTD (HH:MM)	Yellow Colour	Fluorescence
			Total coliforms	Thermotolerant coliforms			
E.coli 1	67	0	46	54	11:00	+	+
K.aerogenes	37	0	37	37	10:48	+	-
E.coli 2	50	0	74	62	10:36	+	+
K.aerogenes	37	0	43	48	10:48	+	-
P.aeruginosa E.coli	240 50	0	91	64	10:42	+	+
P.aeruginosa K.aerogenes	240 37	0	41	35	10:18	+	-
P.aeruginosa E.coli	240 500	0	428	353	9:48	+	+
P.aeruginosa K.aerogenes	240 370	0	270	276	9:48	+	-
E.coli 2	<1	0	1	0	17:24	+	+
K.aerogenes	<1	0	1	1	>18	-	-
E.coli 1	4	0	0	0	>18	-	-
E.coli 1	4	1	0	0	>18	-	-
K.aerogenes	4	0	0	0	>18	-	-
K.aerogenes	4	1	1	0	>18	-	-
K.aerogenes E.coli 2	40 <1	0	51	38	10:48	+	+
K.aerogenes E.coli 2	40 <1	1	34	15	10:48	+	+
K.aerogenes E.coli 2	40 <1	5	0	0	>18	-	-

4 DISCUSSION

Recent interest in coliforms and *Escherichia coli* as bacterial indicators of water quality and treatment efficacy has led to the development of several new culture media for their detection. Most of the novel formulations offer advantages relative to the accepted products in terms of detection specificity, test duration and ease of use: in the present study, examples of these improved media were evaluated using the RABIT system with the objective of further increasing their usefulness by permitting quantitative detection of the target organisms.

Merck Fluorocult media allowed particularly rapid detection of *Escherichia coli* and *Klebsiella aerogenes* in RABIT. An inoculum of <20 cfu of either organism was detected within 10 hours in all cases. Fluorescence and indole production as indicators of *E.coli* presence were found to be reliable and convenient tests. It should be noted, however, that confirmation of *E.coli* by indole production should be performed using media incubated at 44°C: this would necessitate a two-tube test (coliforms and thermotolerant coliforms/*E.coli*) for each sample. The performance of Fluorocult media incubated at 44°C in RABIT was not evaluated.

Of the two Fluorocult formulations tested, BRILA broth offered the advantage that it could be used in the direct RABIT cell, which has a working volume of 10 ml (c.f. 5 ml in the indirect cell). This permits the analysis of larger sample volumes, though both formulations could in theory be used to examine larger volumes of water by passing the samples through membrane filters and incubating these in the RABIT cell.

Both Colilert formulations evaluated in this study generated strong signals from the growth of coliforms and *E.coli* in the indirect RABIT technique. Time-to-Detection for a given inoculum was 3 to 4 hours faster using Colilert 18 than that obtained from the original formulation; however the results obtained from pure cultures suggested that 18h incubation would be adequate for both media in RABIT. The reliability of the colour change and fluorescence production after a reduced incubation period was not investigated, but at the manufacturer's recommended reading time, these indicators worked well in both formulations. The use of ONPG for the detection of coliforms on the basis of ß-galactosidase activity rather than simply lactose fermentation provides a highly specific test. Furthermore, the use of Colilert media in RABIT in conjunction with filtration allows 20 tests to be performed using the manufacturer's single-test pack, representing a considerable cost saving. RABIT results have also been shown to correlate well with membrane filter counts obtained for total coliforms in chlorinated and unchlorinated water samples. The impedance method thus offers a potential alternative to the rather laborious membrane filtration procedure: the simple test initiation routine of the RABIT system combined with the provision of fully automated results both reduces "hands-on" time and removes the subjective errors associated with the enumeration of specific colony types on filter membranes.

In conclusion, the use of Fluorocult and Colilert media in conjunction with an impedance system such as the RABIT could provide a truly automated alternative to traditional methods of enumerating water-borne bacteria, reducing the analysis time and operator input whilst taking full advantage of the specific indicator systems and confirmatory tests provided by these media. However, the use of these media in impedance systems requires more extensive evaluation in order to assess the suitability of the test methods for analysis of a wider range of samples, including naturally contaminated water.

References

Anon, 1994 The Bacteriological Examination of Drinking Water Supplies (Reports on Public Health and Medical Subjects No 71): HMSO.

Chapter 5
The Development And Evaluation Of A New System For The Simultaneous Enumeration Of Coliforms And *E. Coli*

Brian Thomas, Rebecca Williams, Simon Forster and Peter Grant

Celsis Ltd., Cambridge Science Park, Milton Road, Cambridge, CB4 4FX, UK

1 INTRODUCTION

Escherichia coli and other coliforms are used as indicator organisms in water. Their presence is presumed to result from contamination of the water supply and to indicate the risk that the water contains other more harmful but less easily detectable pathogens. The enumeration of these indicator organisms in water samples is a major component of the workload of microbiology laboratories in many industries.

The precise techniques utilised have evolved over the years. The Most Probable Number (MPN) method involves the division of the sample into a number of portions, each of which is incubated in a container of culture medium capable of demonstrating the presence of coliforms, usually by the fermentation of lactose to produce acid and gas, whilst inhibiting the growth of non-target organisms. The number of coliforms present in the original sample is calculated statistically from the number and volume of the sample aliquots yielding positive results. Confirmatory tests may be performed on the positive cultures to detect the presence of *E. coli*. The major disadvantage of this technique is the amount of labour involved in dividing the sample and processing the many cultures.

The Membrane Filtration technique differs in that microbial cells are captured from the sample on a filter which is then incubated on a suitable selective and differential medium in the form of an agar or an absorbent pad soaked in a liquid broth. Coliforms form colonies which can be differentiated from those of non-target organisms and counted. Specific enumeration of *E. coli* is achieved by dividing either the sample or the filter and varying the incubation conditions (usually the temperature) so that only this species is capable of growth. Alternatively, the identity of the coliform colonies may be established by confirmatory tests. Drawbacks of this method include the need to perform separate assays for *E. coli* and coliforms, the difficulty in counting colonies on an often overcrowded filter and the labour involved.

The Celsis Digital System is an attempt to combine the advantages of both of these established techniques. Originally designed for use with ATP bioluminescence as a rapid method of performing total microbial counts, we have further refined the system by the use of chromogenic enzyme substrates for the simultaneous enumeration of both *E. coli* and coliforms in a single sample.

2 DESCRIPTION OF THE DIGITAL SYSTEM

The two basic components of the Celsis Digital System are a 96 well disposable microplate and a 96 chamber reusable sample manifold (Figure 5.1). The wells of the microplate have a 0.22μ pore size membrane filter welded to their base and allow the filtration of aqueous samples. The sample manifold clamps to the top of the microplate. It is divided internally into 4 quadrants, each consisting of 24 chambers corresponding to the wells of the microplate. Each quadrant divides a single sample into 24 aliquots which are then filtered through the corresponding 24 wells of one quadrant of the microplate. Any organisms present are retained by the filters. Various sizes of manifold are available. The one evaluated here allowed four 100 ml samples to be processed simultaneously.

Following filtration, the manifold is disengaged from the microplate and can be autoclaved for reuse. The microplate is placed on an absorbent pad soaked in a suitable culture medium and incubated for 18 to 24 hours after which the proportion of the 24 wells per sample that exhibit growth can be determined. The growth detection method utilised may vary from one application to another. In this case, the use of Colorcount medium causes wells containing *E. coli* or other coliforms to change colour (see below). The most probable number (MPN) of both organisms present in the original sample can be calculated from the number of coloured wells by reference to the MPN table provided.

2.1 Description of digital Colorcount medium

Colorcount medium is a nutrient medium containing inhibitors to suppress the growth of many non- coliform organisms and indicators in the form of chromogenic enzyme substrates to detect the presence of *E. coli* and other coliforms.

Figure 5.1 *Digital Microplate and manifold*

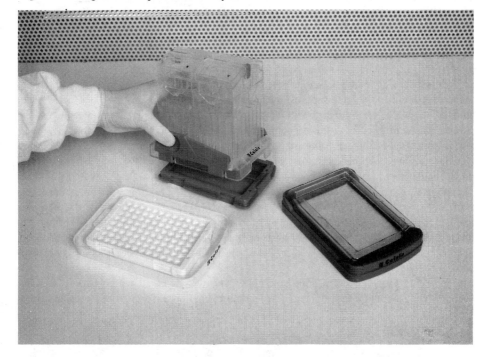

Coliforms are detected by their possession of the enzyme beta- galactosidase. The substrate for this enzyme incorporated in the medium is colourless but yields an insoluble pink chromogen in the presence of coliforms. The insoluble nature of the chromogen ensures that the colour is retained in the well and cannot diffuse to interfere with the interpretation of adjacent wells. A Digital well containing a coliform thus becomes pink whilst wells not containing coliforms remain colourless.

A second substrate in the medium detects the activity of the enzyme beta-glucuronidase which is diagnostic for the presence of *E. coli* . In this case, the colourless substrate yields an insoluble blue colour. Digital wells containing *E. coli* thus become blue or purple (*E. coli* is, of course, a coliform and will posses both enzymes thus generating both pink and blue chromogens leading to a purple endpoint).

Non-coliform organisms that may also produce these enzymes are suppressed by selective agents incorporated into the medium.

2.2 Description of the enumeration principle

The Digital sample processing method divides each 100 ml sample into 24 equal parts and filters each of these through a single microplate well. A well containing one or more cells of the target organism will change colour during the 18 to 24 hour incubation period. As there is a chance that a well will be inoculated with more than one cell, it is not possible to establish the number of target organisms present in the original sample directly from the number of coloured wells. The number can, however, be calculated statistically using the Most Probable Number or MPN technique. In practice this simply involves counting the number of coloured wells and referring to an MPN table supplied with the system.

The range of the system is from 0 to 50 organisms per sample, the upper number being the one where the statistical probability of all 24 wells being positive becomes significant. If a sample is expected to contain sufficient organisms to overload the system because all 24 wells will be coloured then it is possible to dilute in sterile water prior to filtration.

2.3 Evaluation of the system

Evaluations were conducted in our own laboratories and in those of a UK water utility company.

Pure laboratory cultures were used to confirm the ability of the system to detect a variety of species of coliform and to suppress the growth of non-target organisms. The species tested and the number of strains of each are shown in Table 5.1. In our laboratory, the identity of the strains was confirmed with the appropriate API identification system. Standard UK water industry methods (Anon, 1983) were used in the water utility's laboratory.

For the target organisms, experiments consisted of diluting broth cultures of each strain in sterile de-ionised water and adding 100 µl to a single well of a Digital plate, giving an inoculum level of approximately 10 colony forming units (cfu) per well. Inoculum levels were confirmed by culture on tryptone soya agar (TSA). Following filtration, the plate was incubated for 18 hours on an absorbent pad containing 20 ml of CCC medium and, at the end of this period, the colour of each well was recorded. Experiments with non-target organisms were similar except that the cultures were diluted to give a higher inoculum level of approximately 10^4 cfu per well.

The ability of the system to enumerate organisms was tested with a variety of water sample types including those obtained from natural environments such as ponds and rivers

Table 5.1 *Laboratory strains tested*

Species	No. of Strains	Species	No. of Strains
E. coli	53	Bacillus cereus	1
E. coli (wild isolates)	27	Bacillus subtilis	1
Klebsiella pneumoniae	7	Candida albicans	1
Klebsiella ornitholytica	3	Proteus mirabilis	1
Enterobacter agglomerans	3	Providencia stuartii	1
Enterobacter cloacae	4	Proteus vulgaris	1
Enterobacter gergoviae	1	Flavobacterium indologenes	3
Citrobacter freundii	12	Salmonella abony	1
Coliforms (wild isolates)	25	Pseudomonas aeruginosa	10
Aeromonas hydrophila	5	Pseudomonas fluorescens	8
Aeromonas salmonicida	1	Pseudomonas cepacia	3
Aeromonas (wild isolates)	60	Staphylococcus aureus	1
Yersinia enterocolitica	1	Legionella pneumophila	1

and from water treatment systems. Sea water samples were also tested. Samples were divided into 100 ml aliquots and were processed by the Digital Colorcount method and by conventional membrane filtration using membrane lauryl sulphate broth (Anon, 1983). A single aliquot was tested by the Digital Colorcount method but two aliquots were necessary for the membrane filtration method to allow incubation at both 37 (coliforms) and 44 °C (*E. coli*), this being the means of differentiating these organisms using this method.

3 RESULTS AND DISCUSSION

The results of testing pure laboratory cultures in the Digital Colorcount system are shown in Table 5.2. All non-*E. coli* strains of coliform tested except 1 yielded the expected pink colour in the well of a Digital plate. This supports the choice of substrate for the detection of beta-galactosidase in these organisms and the ability of the medium to support the growth of a variety of coliform species.

76 of 80 strains of *E. coli* produced a blue/purple colour in the wells. The remaining 4 strains gave a pink reaction indicating that they had grown in the assay and produced beta-galactosidase but not beta-glucuronidase. The identity of these strains was confirmed as *E. coli* using the API 20E identification test system. It is known that a small percentage of strains of *E. coli* do not exhibit beta-glucuronidase activity including the O-157 serovar. These strains were, however, detected as coliforms.

Table 5.2 *Digital Colorcount results with laboratory strains*

Organism	Colourless	Pink	Blue
E .coli	0	4	76
Other coliforms	1	62	
Non-target organisms	100	0	0

All of the non-target strains gave the expected colourless negative reaction, even at the relatively high inoculum levels tested. A large number of Aeromonas were included in this study because it is known that these organisms produce beta-galactosidase and have given false positive reactions in other coliform assays based on the detection of this enzyme. They are also relatively common in some water distribution systems. The success of the Digital Colorcount assay in overcoming this problem was due to the selective agents employed and to the choice of substrate for the detection of beta-galactosidase activity.

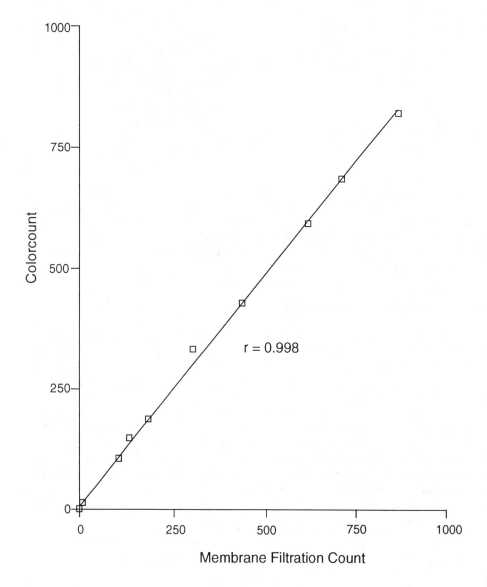

Figure 5.2 *Digital Colorcount enumeration of E. coli compared with membrane filtration counts*

The results of enumerating of *E. coli* and coliforms in natural water samples in comparison with membrane filtration colony counts obtained on MLSB are shown in Figures 5.2 and 5.3 respectively. Correlation coefficients of 0.99 for both types of organism indicate that the two methods produce equivalent results.

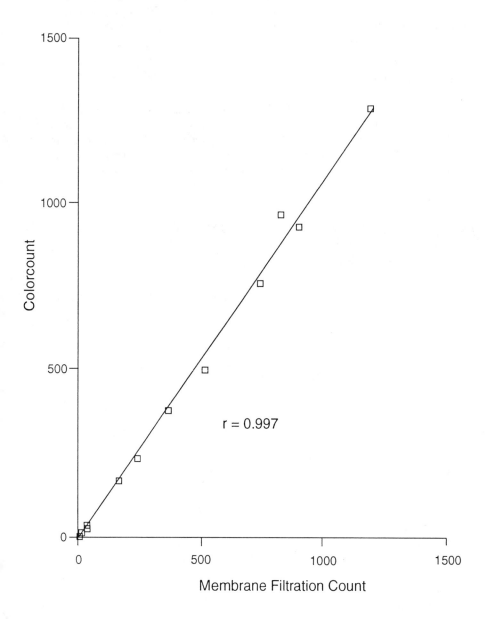

Figure 5.3 *Digital Colorcount enumeration of coliforms compared with membrane filtration counts*

A major feature of the system was the reduction in labour and materials employed compared to the conventional method. The processing of up to 4 samples on a single microplate together with the fact that *E. coli* and coliforms are enumerated simultaneously meant that a single Colourcount assay was equivalent to 8 conventional tests. The simplicity of interpretation compared to standard colony counting on membrane filters, the convenience of a ready prepared liquid medium and ease of sub-culture for confirmatory tests further contributed to the practicability of the system.

4 CONCLUSIONS

The studies reported here demonstrate that the Digital Colorcount system is capable of detecting and enumerating both *E. coli* and other coliforms simultaneously in a single water sample. The results obtained are equivalent to those of the conventional membrane filtration method whilst considerable labour and material savings are possible. Further studies are now underway to establish the practicability of the Digital Colorcount system in other industries.

References

Anon. The Bacteriological Examination of Drinking Water Supplies 1982. Reports on Public Health and Medical Subjects No. 71. London: HMSO, 1983.

Coliforms and Food

Chapter 6
The Significance Of Coliforms *Escherichia Coli* And The Enterobacteriaceae In Raw And Processed Foods

C L Baylis & S B Petitt

Department of Food Microbiology, Leatherhead Food Research Association, Randalls Road, Leatherhead, Surrey KT22 7RY.

1 INTRODUCTION

Specific groups of bacteria or individual species are commonly used to provide evidence of poor sanitary practices, inadequate processing or post-process contamination of foods. These organisms are commonly referred to as indicator organisms, the most popular of which is the coliform group. Particular groups or genera may also be used to assess the potential risk of closely related pathogens being present in foods, a function for which Ingram proposed the term index organism (Mossel, 1978; Mossel, 1982).

The most popular indicator group has been the coliform group, although, as a whole, this group does not have great specificity. Indeed, coliforms have been defined as microorganisms that give a positive result to a coliform test (Anon, 1978). The genera detected by the coliform test are those members of the Enterobacteriaceae that are able to decompose lactose with acid and gas production, namely *Escherichia, Arizona, Citrobacter, Erwinia, Enterobacter* and *Klebsiella*. Slow lactose-fermenting or lactose-negative Enterobacteriaceae, including the enteric pathogens *Salmonella* and *Shigella* and those lactose-negative *E. coli* (for example *Paracolobactrum coliforme*) that exist (Mossel, 1957; Buttiaux, 1959; Buttiaux and Mossel, 1961) would be overlooked by restricting enumeration to coliform organisms. For this reason, foods are often examined for total Enterobacteriaceae instead, which are a taxonomically well defined group. Furthermore, this eliminates the continuing question of what is the definition of a coliform and which genera belong to this ill-defined yet commonly used group of organisms.

Although certain coliforms are the normal inhabitants of the intestinal tract of humans, warm-blooded animals and birds, others may also be associated with the soil or plants. Consequently, the presence of coliform organisms in certain raw foods, such as vegetables, may be unavoidable and would not necessarily indicate poor hygiene or unacceptable quality (Splittstoesser, 1983). For example, *Erwinia* species in some foods of plant origin may be tolerated, but the presence of *E. coli* could indicate faecal contamination and the potential presence of other enteric pathogens, which is clearly unacceptable. In this situation, it is necessary to identify the species present when making the interpretation.

In foods treated by heat, the coliform test is a useful means of assessing evidence of inadequate processing, poor sanitary practices or post-process contamination. In this situation the actual species present is not important. However, examination of foodstuffs for lactose-positive bacteria only may lead to dangerously false conclusions. This was demonstrated by

Buttiaux and Mossel (1961), who showed that a sample of egg white solids could be passed by a laboratory on the basis of a negative coliform test, despite the fact that it contained 10^4 salmonellae per g^{-1}.

To improve the situation, with respect to assessment of food safety and quality, it is now common practice to use a total Enterobacteriaceae count, instead of limiting the search to coliforms. This reduces the risk of samples contaminated with only lactose-negative members of the group being passed without further investigation. The search for indicator organisms must, however, be in addition to a direct search for enteric pathogens. This is because of the variations in proportions that exist between *Salmonella* and Enterobacteriaceae in various processed foods (Drion and Mossel, 1977). However, because enteric pathogens tend to be present at low levels and distributed in a highly heterogeneous manner throughout a sample, indicator organisms remain a useful tool for the food microbiologist.

A more useful means of assessing faecal contamination of foods is to use the faecal coliforms, particularly *E. coli* as the indicator. The objective is to detect coliforms of faecal origin, on the basis of their ability to ferment lactose and produce gas at 44.5 ± 0.2 °C within 24 ± 2 h (Ray, 1989). However, although the tests for faecal coliforms may provide better specificity for assessing faecal contamination, some non-faecal coliforms are also capable of lactose fermentation at these temperatures. The term thermotolerant coliforms is therefore considered to be a more appropriate description for these organisms. In processed foods, however, there are likely to be cells that have sustained some form of sub-lethal injury. Incubation temperatures above 40 °C could therefore be lethal to the injured indicator bacteria (Ray, 1989) while others may be anaerogenic, leading to the false assumption, on the basis of test results, that the food was free from detectable levels of these bacteria or that acceptable levels were present.

In addition to its role as an indicator of faecal contamination, *E. coli* is one of the most commonly used index organisms currently employed to assess the potential risk of enteric pathogens being present in food. It is clear that certain organisms may therefore serve as both an indicator and index even in the same food (Mossel, 1982), although in both cases a clear distinction between raw and processed foods must be made.

Despite its popularity as an index organism for salmonellae in foods and as an indicator of faecal contamination, *E. coli* does have some limitations. For instance *E. coli* and other coliforms may actually be absent from the gut of certain animals or less abundant than other organisms that inhabit the intestine (Catsaras, 1995). Furthermore, *E. coli* is generally not as resistant as the enteric pathogens, particularly *Salmonella* that it is intended to indicate. It is killed by pasteurisation and rapidly dies off during storage under conditions of drying, freezing and refrigeration, and at low pH (Ray, 1989).

In the environment, whereas most pathogens persist, *E. coli* often disappears. In sewage, for example, survival of *E. coli* is short, whereas *Salmonella paratyphi* B and most other enteritis-salmonellae are able to survive longer and under certain circumstances even proliferate (Müller and Mossel, 1982). A further shortcoming of using *E. coli* as an indicator bacterium is that it does not survive well in the marine environment and is therefore a poor indicator of non-enteric pathogens and pathogens of marine origin.

Unlike the situation in water, where indicator bacteria have little or no chance of multiplying, in certain foods these organisms may continue to increase in number. Even at refrigeration temperatures some members of the Enterobacteriaceae may continue to grow (Mossel and Zwart, 1960). A study by Panes and Thomas (1959) reported multiplication of these organisms in milk held at 3-5°C. In chilled meats, Eddy and Kitchell (1959) reported growth of twenty-eight strains of coli-aerogenes bacteria at + 1·5 °C and some at -1·5 °C. Further investigations revealed that the majority of these strains had optimum growth temperatures of 37°C whilst for the remainder it was 30°C. Therefore, they are not true psychrotrophs and it was suggested that these organisms would be more accurately described as cold-tolerant mesophiles.

The ability of these organisms to multiply in certain foods makes interpretation of enumeration results difficult as the numbers present may not necessarily reflect the degree of initial contamination. In certain foods, such as refrigerated raw meat, both the Enterobacteriaceae and the coliforms may be considered to be of little value as indicators of faecal contamination. This is partly because of the ability of the so-called psychotrophic strains to multiply and also because these groups include organisms which are not of faecal origin (Shaw and Roberts, 1982). In contrast, *E. coli* is generally of faecal origin and growth of this organism under refrigeration conditions is minimal; thus it may still be a satisfactory indicator of faecal contamination in refrigerated raw meat (Shaw and Roberts, 1982; Harris and Stiles, 1992) and also in fresh poultry stored at 6 °C (Zeitoun, Debevere and Mossel, 1994).

In most so-called perishable foods, those with a pH of over 4·5, faecal indicator organisms will continue to multiply actively, unless unfavourable conditions prevail (Buttiaux and Mossel, 1961). Their presence at significant levels may therefore indicate faulty storage conditions or temperature abuse, although care should be taken when interpreting such results because of the multiplication of so-called psychrotrophic Enterobacteriaceae. Multiplication of faecal coliforms can also occur on food-contact surfaces and equipment so the assessment of sanitary conditions should also be interpreted with care. The same prudence is necessary when applying microbiological standards or specifications to a food that are ultimately used as a means of assessing the microbiological quality or safety of that food. Criteria therefore, should be realistic. They must be based on accurate knowledge of microbial levels encountered in foods determined from data obtained using proven reproducible methods and from carefully designed surveys (Baird-Parker, 1976).

In some foods, such as seafood, the usefulness of the Enterobacteriaceae, in particular the coliform group, to indicate the presence of non-enteric pathogens, viruses or psychrotrophic pathogens remains questionable, largely owing to the different nutritional and environmental requirements of these organisms. In a study comparing the relationship between classical indicator bacteria and pathogens involved in shellfish-borne diseases (Martinez-Manzanares *et al.*, 1991) no significant correlation could be found. Total coliforms and *E. coli* were, however, thought to be the best indicators of the presence of pathogens at low and high concentration intervals, respectively. In another study (Gilbert, 1982), no correlation was found between the isolation of *Salmonella* and the presence of *E. coli* or high colony plate counts in some cooked seafoods, which further highlights the limitations that may exist when using *E. coli* as an index for certain foods.

The situation with respect to *E. coli* has also changed dramatically with the emergence of enteropathogenic, enteroinvasive and enterotoxigenic strains. Unlike the normal *E. coli* strains that colonise the human bowel and remain as harmless commensals, these pathogenic strains are the cause of distinct diarrhoeal disease and foodborne illness (Jay, 1992). More recently we have seen the emergence of the so-called enterohaemorrhagic strains or verocytotoxigenic *E. coli* (VTEC). These are often represented by *E. coli* O157:H7 (Padhye and Doyle, 1992), which was recognised as a foodborne pathogen in 1982. The mode of transmission is primarily through food, of which there have been numerous documented outbreaks. To date foods implicated with foodborne transmission of *E. coli* O157 include ground beef products, such as hamburgers and beefburgers, apple cider and untreated and contaminated milk.

Unlike normal *E. coli* strains, however, growth of *E. coli* O157:H7 is poor at 44 to 45·5°C (Doyle and Schoeni, 1984), which is the normal temperature range used to detect *E. coli* and other faecal coliforms. Biochemical differences also distinguish them from conventional *E. coli*, namely, their inability to ferment sorbitol and the absence of β-glucuronidase activity. Detection at present relies upon specific isolation methods and confirmation techniques. However, the similarity between normal *E. coli* and some enterohaemorrhagic strains makes detection and isolation particularly difficult.

This was highlighted by a recent outbreak of haemolytic uraemic syndrome in Australia which was discussed at the 8th Australian, Food Microbiology Conference (Buckle, 1995). The cause was enterohaemorrhagic *E. coli* O111 in contaminated Mettwurst sausage. Strains isolated from patients were sorbitol-positive and could not be differentiated culturally from normal *E. coli*. The use of *Salmonella* as an indicator of these strains in such fermented sausages, albeit an unusual yet interesting concept, was discussed at the meeting. Unlike normal *E. coli*, pathogenic strains are not employed as indicators but are themselves the subject of investigation. At present no suitable indicator or index organisms have been found for these strains, although the use of suitable coliphages is currently under investigation.

The introduction of new technology, particularly immunological techniques and improvements to conventional enrichment methods have greatly improved the degree of reliability and effectiveness of detecting enteric pathogens in foods. Nevertheless, in addition to the direct search for pathogens, the examination of foods for suitable indicator organisms remains important, despite all the improvements in detection methods. In certain foods, however, the search for more suitable indicator and index organisms continues.

Firstly, owing to the highly heterogeneous distribution of these bacteria, failure to detect pathogenic Enterobacteriaceae is of limited significance. Secondly, the detection of certain pathogens may only be entrusted to a specialist laboratory and, thirdly, the probability of enteric pathogens being present in a food commodity is extremely low when tests on a series of samples have shown that indicator organisms are absent. Absence of enteric pathogens however, has significance only to the sample drawn from a given consignment of food (Drion and Mossel, 1977).

In addition to the Enterobacteriaceae, the *enterococci* or the *faecal streptococci* group is sometimes used as an indicator of faecal contamination of foods (Buttiaux, 1959; Buttiaux and Mossel 1961; Mossel, 1978, Jay, 1992; Catsaras, 1995). However, although they are more resistant to heat and can survive better than coliforms in refrigerated, frozen, dried and low pH environments (Ray, 1989; Catsaras, 1995), they too have similar limitations, namely their ability to grow on equipment surfaces and to multiply in foods. Other organisms that have been used as indicators of faecal contamination include sulfite reducing *Clostridium* species, although it is the Enterobacteriaceae and coliforms that remain the most popular indicator organisms.

References

Anon Microorganisms in Foods. Vol. 1 - Their Significance and Methods of Enumeration". International Commission on Microbiological Specifications for Foods. 2nd Edition. University of Toronto Press, Toronto, 1978.

Baird-Parker AC. Microbiological standards for foods. Fleischwirtschaft 1976; 56: 96

Buckle K. Conference Report 8th Australian food microbiology conference. Trends in Food

Science & Technology 1995; May (6): 163-65.

Buttiaux R. The value of the association Escherichiae - group D streptococci in the diagnosis of contamination in foods. Journal of Applied Bacteriology 1959; 22 (1): 153-8.

Buttiaux R, Mossel DAA. The significance of various organisms of faecal origin in foods and drinking water. Journal of Applied Bacteriology 1961; 24 (3): 353-64.

Catsaras MV. Indicators of fecal contamination. In: Bourgeois CM, Leveau JY., editors, Microbiological control for foods and agricultural products. Cambridge: VCH Publishers, 1995: 293-308.

Doyle MP, Schoeni JL. Survival and growth of *Escherichia coli* associated with hemorrhagic colitis. Applied and Environmental Microbiology 1984 Oct; 48 (4): 855-56.

Drion EF, Mossel DAA. The reliability of the examination of foods, processed for safety, for enteric pathogens and Enterobacteriaceae: a mathematical and ecological study. Journal of Hygiene 1977; 78: 301-24.

Eddy BP, Kitchell AG. Cold-tolerant fermentative gram-negative organisms from meat and other sources. Journal of Applied Bacteriology 1959; 22 (1): 57-63.

Gilbert RJ. Indicator organisms and fish and shellfish poisoning. Antonie van Leeuwenhoek 1982; 48: 623-5.

Harris LJ, Stiles ME. Reliability of *Escherichia coli* counts for vacuum-packed ground beef. Journal of Food Protection 1992; 55, (4): 266-70.

Jay JM.Modern food microbiology. 4th ed. New York: Nostrand Reinhold, 1992.

Martinez-Manzanares E, Morinigo MA, Cornax R, Egea F, Borrego JJ. Relationship between classical indicators and several pathogenic microorganisms involved in shellfish-borne diseases. Journal of Food Protection 1991; 54 (9): 711-7.

Mossel DAA. The presumptive enumeration of lactose negative as well as lactose positive Enterobacteriaceae in foods. Applied Microbiology 1957; 5: 379-81.

Mossel DAA. Index and indicator organisms - a current assessment of their usefulness and significance. Food Technology in Australia 1978 Jun; 30 (6): 212-9.

Mossel DAA. Marker (index and indicator) organisms in food and drinking water. Semantics, ecology, taxonomy and enumeration. Antonie van Leeuwenhoek 1982; 48: 641-4.

Mossel DAA, Zwart H. The rapid tentative recognition of psychrotrophic types among Enterobacteriaceae isolated from foods. Journal of Applied Bacteriology 1960; 23 (2): 185-8.

Müller HE, Mossel DAA. Observations on the occurrence of *Salmonella* in 20 000 samples of drinking water and a note on the choice of marker organisms for monitoring water supplies. Antonie van Leeuwenhoek 1982; 48: 641-4.

Panes JJ, Thomas SB. The multiplication of coli-aerogenes bacteria in milk stored at 3-5°. Journal of Applied Bacteriology 1959; 22 (1): 57-63

Padhye NV, Doyle MP. *Escherichia coli* O157:H7: Epimediology, pathogenisis, and methods for detection in food. Journal of Food Protection, 1992; 55 (7), 555-65.

Ray B. Injured index and pathogenic bacteria: occurrence and detection in foods, water and feeds. Boca Raton, FL:CRC Press, 1989.

Shaw BG, Roberts TA. Indicator organisms in raw meats. Antonie van Leeuwenhoek 1982; 48: 641-4.

Splittstoesser DF. Indicator organisms on frozen blanched vegetables. Food Technology 1983: 37 (6): 105-6.

Zeitoun AAM, Debevere JM, Mossel DAA. Significance of Enterobacteriaceae as index organisms for hygiene of fresh untreated poultry, poultry treated with lactic acid and poultry stored under modified atmosphere. Food Microbiology 1994; 11: 169-76.

Chapter 7
Some Observations On The Ecology Of *Escherichia Coli* Isolated From Healthy Farm Animals

M Hinton and V M Allen

Division of Food Animal Science, University of Bristol, Langford House, Langford, Avon BS18 7DU

1 INTRODUCTION

Escherichia coli is a heterogeneous species comprising many different strains, the vast majority of which are not pathogenic (Hinton, 1985). The biochemical characteristics and the classification of the species have been reviewed recently (Bettleheim 1994; Lior 1994). Strains can be identified using serotyping, biochemical typing, resistogram typing, plasmid profiling etc (Hinton, 1985) and continued interest in the differentiation of isolates is reflected in recent papers on biotyping (Crichton and Taylor 1995) , plasmid profiling (Al-Sam *et al.*, 1993; O'Brien *et al.*, 1993), resistogram typing (David *et al.*, 1992) and ribotyping (Tarkka *et al.*, 1994).

There have been many studies on the ecology of both commensal and pathogenic *E. coli* in the farm environment (Hinton *et al.*, 1984; 1985a,b, 1986; Hinton, 1986; Linton and Hinton, 1988). In contrast, there have been few detailed studies of the ecology of *E. coli* in abattoirs and boning plants.

Pathogenic strains may be associated with enteric disease in new born farm animals while *E. coli* O157:H7 and other verotoxic *E. coli* (VTEC) strains, may be responsible for haemorrhagic colitis and the haemolytic uraemic syndrome in humans. The pathogenic aspect of *E. coli* has been the focus of much attention over the years and many aspects of pathogenesis are covered in books such as those edited by Sussman (1985) and Gyles (1994).

This paper comprises two principal parts which consider, respectively, some aspects of the ecology of *E. coli* in (1) healthy farm animals and (2) on meat and in the abattoir. Other foods of animal and plant origin are also recognised sources of *E. coli* for the human population but they are outwith the scope of this review.

2 *ESCHERICHIA COLI* IN FARM ANIMALS

The intestinal tract of newborn animals becomes colonised with *E. coli* within a matter of hours of birth . This population is unstable and there is a rapid turn over of strains (biotypes) comprising the 'majority' flora such that, with daily sampling of faeces, most of the strains are isolated on only one occasion. The strains are probably 'picked up' from the environment and as a consequence the prevalence of resistance in the intestinal *E. coli* reflects that present in the accommodation in which the animal is housed (Hinton *et al.*, 1984, 1985a,b, 1994). The *E. coli* population is more complex, and the prevalence of resistance tends to be higher, in younger animals than in those that are mature.

The administration of antibiotics at therapeutic concentrations, and even antibiotics secreted in the milk, may select rapidly for resistant non-pathogenic *E. coli* strains as long as they are present for selection in the intestinal microbiota (Lim 1985; Hinton and Linton, 1986; Hinton *et al.*, 1987). Similarly, dosing with antibiotics at below the minimum inhibitory concentration may also modulate the *E. coli* flora by favouring the persistence of resistant strains (Al-Sam *et al.*, 1993).

Resistance to a relatively newly introduced antibiotic, such as apramycin, can become widely disseminated in non-pathogenic *E. coli* strains isolated from both animals and the environment (Hunter *et al.*, 1994), while people working on farms and in abattoirs may have a more resistant *E. coli* flora than those resident in (sub)urban communities (Nijsten *et al.*, 1994). Once present on a farm, resistant enteric bacteria (e.g. coliforms) may persist in the animal population for many years after the discontinuation of the use of all antibiotics (Langlois *et al.*, 1988).

That cattle can be a potential source of VTEC for humans has been demonstrated both in Europe and North America (e.g. Chapman *et al.*, 1993; Hancock *et al.*, 1994).

3 *ESCHERICHIA COLI* IN ABATTOIRS AND ON MEAT

Many investigations of abattoir hygiene have involved the enumeration of coliforms, rather than *E. coli* (e.g. Roberts *et al.*, 1980b; Whelehan *et al.*, 1986; Charlebois *et al.*, 1991). Coliforms should not, however, be considered a reliable indicator of the presence of *E. coli* (Newton 1979), since this group of bacteria lactose fermenting comprises a number of genera including *Citrobacter*, *Enterobacter*, *Escherichia*, *Serratia* and *Klebsiella*. For instance, in a recent survey coliforms were isolated from all of 102 samples of frozen minced beef. On the other hand, *E. coli* was cultured from only 79 samples and in each case it comprised <10% of the total coliforms (Hinton, unpublished observations).

The contamination of beef and sheep carcasses with either coliforms or *E. coli*, as with any other bacterial group, varies between abattoirs both qualitatively and quantitatively. Many carcasses appear not to be contaminated with either category of organism, while, for those that are, the numbers on the surface are typically low i.e. $<10^3$ cfu cm^{-2} (e.g. USDA & FSIS, 1994; Hadley, 1995; Hudson & Hinton, unpublished observations).

Important sources of *E. coli* on carcasses are presumed to be the faeces, contents of the intestinal tract or fleece, hide, skin. The importance of the fleece has been illustrated by Hadley (1995) who found that the prevalence of contamination of sheep carcasses with coliforms, and the median number, increased with increasing faecal contamination of the fleece covering abdomen, legs.

Pig carcasses may remain contaminated after 'scalding', dehairing and 'polishing' (Gill 1994); while there is evidence that on pig and sheep carcasses strains of *E. coli* can be removed by washing and replaced by others present in the environment (Linton *et al.*, 1977a; Bettleheim 1981).

The organs and muscle tissue may become contaminated with *E. coli* and other organisms placed on the captive bolt, sticking knife and pithing rod (Mackey and Derrick 1979). An investigation in a poultry processing plant indicated that *E. coli* strains did not persist on equipment, however, and were continually replaced (Cherrington *et al.*, 1988). This is in contrast to certain strains of *Staphylococcus aureus* which may colonise equipment in poultry processing plants for many months (e.g. Dodd *et al.* 1988).

The relative importance of the faeces, contents of the intestinal tract, outer surface, equipment and the environment as sources of *E. coli* on carcass meat is not known. Within abattoirs some sites on the carcass surface are consistently more heavily contaminated with coliforms and these sites may differ between abattoirs or contamination may vary at the same sites at different visits to the same abattoir (Roberts *et al.*, 1980b; Hudson *et al.*, 1987). The numbers of coliforms may not be influenced specifically by dressing, trimming or chilled storage although the numbers may increase during cutting operations (Charlebois *et al.*, 1991).

The proportion of warm carcasses and chilled meat samples contaminated with *E. coli* may differ in the different mammalian species. For instance, McCulloch and Whitehead (1981) isolated *E. coli* frequently from the surface of recently killed pigs and only sporadically from cattle and sheep while Bensink and Bothmann (1991) reported that 13%, 24% and 29% of samples of beef, lamb and pork meat samples were contaminated with the proportion of multiply resistant isolates being greatest on pork.

The microbiological quality of minced (ground) beef from retail outlets in the USA and UK has been investigated (e.g. Foster *et al.*, 1977; Roberts *et al.*, 1980a; Hudson *et al.*, 1986). Coliforms and *E. coli* were found in most samples although the numbers were generally low ($< 10^3$ cfu g^{-1}). A more recent survey of retail products in South Africa indicated 75%, 79% and 28% of minced beef, broiler carcasses and processed meats (Vienna sausage, shoulder ham and cervolat) were contaminated with *E. coli* and that there was no relationship between the total bacterial count and the presence of either *E. coli* or *Staph. aureus* (Vorster *et al.*, 1994).

The number of bacteria on the surface of carcasses, including pathogens such as *E. coli* O157:H7, can be reduced, but not necessarily eliminated, by a number of decontamination procedures including spraying or dipping in organic acids (Biemuller *et al.*, 1973; Cherrington *et al.*, 1991; Brackett *et al.*, 1994; Hardin *et al.*, 1995; Smulders, 1995), trisodium phosphate (e.g Gudmunsdóttir *et al.*, 1993) and hot water and steam at atmospheric pressure (Biemuller *et al.*, 1973; Graham *et al.*, 1978; Smith and Graham, 1978; Davey and Smith, 1989; Davey, 1990). None of these methods are without their drawbacks. For instance, chemical compounds may effect the organoleptic properties of meat if used at high concentrations (see Cherrington *et al.*, 1991) while hot water and steam may produce changes in the appearance of the carcass which may be deemed unacceptable (e.g. Biemuller *et al.*, 1973). It must be borne in mind that the process of decontamination may only injure the organism and that viable cells may persist on the meat but these will not be recoverable if selective isolation media are used. This problem, which may provide a false sense of security, has been investigated in relation to thermally-injured *E. coli* O157:H7 (Ahmed and Conner, 1995).

4 CONCLUSIONS

1. *E. coli* is a common bacterial species which has been studied in great detail in both healthy and diseased animals, including man. Despite this attention much still remains to be learnt about its ecology, particularly when it is inhabiting the environment and not associated with an animal.

Methods have been developed which allow specific pathogenic strains to be identified with certainty, and these, together with techniques which can detect these strains even when present in very small numbers, will assist in the understanding of the ecology of the species *E. coli* as whole. However, it remains imperative that the behaviour of the non-pathogenic strains on farms, and in abattoirs and food processing environments is elucidated since, if there are major differences between pathogenic and non-pathogenic strains, different protocols will be required for the control of contamination of food.

2. The intestinal *E. coli* flora of animals change with time with the turnover of strains tending to be more rapid in the young animal. The strains forming the majority flora may reflect those present in the environment, and possibly feed, and as a consequence the prevalence of resistance to antibacterial agents can increase in some circumstances when there is no antibiotic selection pressure. This phenomenon is of importance since it must be kept in mind when the prevalence of antibiotic resistance is being investigated.

3. Many studies of abattoir hygiene have involved the enumeration of coliforms rather than *E. coli* . This means that the results can only be taken as a guide since coliforms are not reliable indicators of faecal contamination.

4. *E. coli* can be cultured sporadically for the surface of carcasses, usually in low numbers ($<10^3$ cfu cm$^-$). The organisms may be derived from the intestinal tract, body surface, equipment or the abattoir environment although the contribution that each of these may make to the final contamination of the carcass has yet to be quantified. This information will only be obtained by a thorough study of the various strains present in the different places in the abattoir and on the carcass at different points of the production process.

5. That carcass meat can be a source of *E. coli* for people handling the raw meat was proved conclusively by Linton *et al.* (1977b). Epidemiological evidence indicates that meat can assist in the dissemination of VTEC strains from farm animals, particularly, to the human population. Verotoxic *E. coli* survive well in the frozen state and may be isolated from beef carcasses and minced beef and pork although the prevalence is usually low (e.g. Read *et al.*, 1990; Willshaw *et al.*, 1993).

6. It is unlikely that it will be possible to either (1) prevent completely carcass contamination with *E. coli* or (2) eradicate pathogenic *E. coli* from the cattle population. Decontamination procedures may reduce the numbers of *E. coli* and other bacteria on carcass meat but they will not be eliminated and may only be sub-lethally injured. As a consequence, it will remain essential that every care is taken to prevent cross-contamination at all stages of the preparation of food for human consumption.

References

Ahmed NM, Conner DE. Evaluation of various media for recovery of thermally-injured *Escherichia coli* O157:H7. J Food Protect 1995; 58:357-60.

Al-Sam S, Linton AH, Bennett PM, Hinton M. Effects of low concentrations of ampicillin in feed on the intestinal *Escherichia coli* of chicks. J Appl Bacteriol 1993; 75:108-112.

Bensink JC, Bothmann FP. Antibiotic-resistant *Escherichia coli* isolated from chilled meat at retail outlets. N Z Vet J 1991;39:126-8.

Bettelheim KA. The isolation of *Escherichia coli* from a sheep slaughtering line in a abattoir. Comp Immunol Microbiol 1981;4:93-100.

Bettleheim KA, Biochemical characteristics of *Escherichia coli*. In: Giles CL, editor. *Escherichia coli* in domestic animals and humans. Wallingford: CAB International, 1994:3-30.

Biemuller GW. Carpenter JA, Reynolds AE. Reduction of bacteria on pork carcasses. J Food Sci 1973;38:261-3.

Brackett RE. Hao YY, Doyle MP. Ineffectiveness of hot acid sprays to decontaminate *Escherichia coli* O157:H7 on beef. J Food Protect 1994;57:193-203.

Chapman PA, Siddons CA, Wright D , Norman P, Fox J, Crick E. Cattle as a possible source of verocytoxin-producing *Escherichia coli* O157 infections in man. Epidemiol Infect 1993;111:439-47.

Charlebois R, Trudel R, Messier S. Surface contamination of beef carcasses by fecal coliforms. J Food Protect 1991;54:950-6.

Cherrington, C.A., Board, RG. & Hinton, M. 1988 Persistence of Escherichia coli in a poultry processing plant. Letters in Applied Microbiology 1988;7, 141-143.

Cherrington CA, Hinton M, Mead GC, Chopra I. Organic acids: chemistry, antibacterial activity and practical applications. In: Rose AH, Tempest DW, editors. Adv Microbial Physiol 1991;32:87-108.

Crichton, PB, Taylor A. Biotyping of *E. coli* in microwell plates. British Journal of Biomed Sci 1995;52:173-7.

David BB, Purushothaman V, Venkatesan RA. Resistogram typing as an epidemiological tool for *Escherichia coli* isolates of poultry origin. Lett Appl Microbiol 1992;15;96-9.

Davey KR. A model for the hot water decontamination of sides of beef in a novel cabinet based on laboratory data. Int J Food Sci Technol 1990;25:88-97.

Davey KR, Smith MG. A laboratory evaluation of novel hot water cabinet for decontamination of sides of beef. Int J Food Sci Technol 1989;24:305-16.

Dodd CER, Mead GC, Waite WM. Detection of the site of contamination by *Staphylococcus aureus* within the defeathering machinery of a poultry processing plant. Lett Appl Microbiol 1988;7:63-6.

Foster JF, Fowler JL, Ladiges WC. A bacteriological survey of raw ground beef. J Food Protect 1977:40:790-4.

Gill CO. Contamination of pork with bacteria during carcass dressing and breaking. Meat Focus Int November 1994:451-3.

Gyles CL, editor. *Escherichia coli* in domestic animals and humans. Wallingford: CAB International, 1994.

Graham A, Cain BP, Eustace IJ. An enclosed hot water spray for improved hygiene of carcass meat. Cannon Hill (Queensland): Commonwealth Scientific and Industrial Research Organisation, Australia; 1978 Meat Research Report 11/78.

Gudmunsdöttir KB, Marin ML, Allen VM, Corry JEL, Hinton M (1993) The antibacterial activity of inorganic phosphates. In: Löpfe J, Kan CA, Mulder RWAW, editors. Prevention and Control of Pathogenic Micro-organisms in Poultry and Poultry Meat Processing. 11. Contamination with pathogens in relation to processing and marketing. Beekbergen: Agricultural Research Service (DLO-NL), 1993:95-100.

Hadley PJ, An investigation into the relationship between levels of fleece contamination in sheep and carcass microbiology [dissertation]. Bristol: Univ of Bristol, 1995.

Hancock DD, Besser TE, Kinsel ML, Tarr PI, Rice DH, Paros MG. The prevalence of *Escherichia coli* O157:H7 in dairy beef cattle in Washington State. Epidemiol Infect 1994;113:199-207.

Hardin MD, Acuff GR, Lucia LM, Oman JS, Savell JW. Comparison of methods for decontamination from beef carcass surfaces. J Food Prot 1995;58:368-74.

Hinton M. The subspecific differentiation of *Escherichia coli* with particular reference to ecological studies in young animals including man. J Hyg 1985;95:595-609.

Hinton M. The ecology of *Escherichia coli* in animals including man with particular reference to drug resistance. Vet Rec 1986;119;420-6

Hinton M, Allen VM, Linton AH. The effect of management of calves on the prevalence of antibiotic-resistance strains of *Escherichia coli* in their faeces. Lett Appl Microbiol 1994;19:197-200.

Hinton M, Lim SK, Linton AH. The influence of antibacterial agents on the complexity of the *Escherichia coli* flora of chickens. FEMS Microbiol Lett 1987;41:169-73.

Hinton M, Hedges AJ, Linton AH. The ecology of *Escherichia coli* in market calves fed a milk substitute diet. J Appl Bact 1985a;58:27-35.

Hinton M, Kaukas A, Linton AH. The ecology of drug resistance in enteric bacteria. J Appl Bact 1986;61 Symposium Suppl:77S-92S.

Hinton M, Linton AH. The selection of trimethoprim resistant *Escherichia coli* in the faecal flora of piglets sucking a sow treated parenterally with trimethoprim. Lett Appl Microbiol 1986;2:107-10.

Hinton M, Linton AH, Hedges AJ. The ecology of *Escherichia coli* in calves reared as dairy-cow replacements. J Appl Bacteriol 1985b;58:27-35.

Hinton M, Rixson PD, Allen V, Linton AH. The persistence of drug-resistant *Escherichia coli* strains in the majority faecal flora of calves. J Hyg 1984;93;547-57.

Hudson WR, Roberts TA, Crosland AR, Casey JC. The bacteriological quality, fat and collagen content of minced beef at retail level. Meat Sci 1986;17;139-52.

Hudson WR, Roberts TA, Whelehan OP. Bacteriological status of beef carcasses at a commercial abattoir before and after slaughterline improvements. J Hyg 1987;98:81-6.

Hunter JEB, Bennett M, Hart CA, Shelly JC, Walton JR. Apramycin-resistant *Escherichia coli* isolated from pigs and a stockman. Epidemiol Infect 1994;112:473-80.

Langlois BE, Dawson KA, Leak I, Aaron DK. Antibiotic resistance of fecal coliforms from pigs in a herd not exposed to antimicrobial agents for 126 months. Vet Microbiol 1988;18:147-53.

Lim SK. The influence of antibacterial drugs on the *Escherichia coli* population of farm animals [dissertation]. Bristol: University of Bristol, 1985.

Linton AH, Handley B, Osborne AD, Shaw BG, Roberts TA, Hudson WR. Contamination of pig carcasses at two abattoirs be *Escherichia coli* with special reference to O-serotypes and antibiotic resistance. J Appl Bacteriol 1977a;42:89-110.

Linton AH, Howe K, Bennett PM, Richmond MH, Whiteside EJ. Colonization of human gut by resistant *Escherichia coli* from chickens. J Appl Bacteriol 1977b;42:365-78.

Linton AH, Hinton M. Enterobacteriaceae associated with animals in health and disease. J Appl Bacteriol 1988;65 Symposium Suppl: 71S-85S.

Lior H. Classification of *Escherichia coli*. In: Giles CL, editor. *Escherichia coli* in domestic animals and humans. Wallingford: CAB International, 1994:31-72.

Mackey BM, Derrick CM. Contamination of deep tissues of carcasses by bacteria present on the slaughter instruments or in the gut. J Appl Bact 1979;46:355-66.

McCulloch B, Whitehead CJ. Bacterial contamination of warm carcass surfaces in relation of total aerobic and coliform counts to the recovery of *Escherichia coli*. J S Afr Vet Assoc 1981;52:119-22.

Newton KG. Value of coliform tests for assessing meat quality. J Appl Bact 1979;47:303-7.

Nijsten R, London N, van den Bogaard A, Stobberingh E. Resistance in faecal *Escherichia coli* isolated from pigfarmers and abattoir workers. Epidemiol. Infect. 1994;113:45-52.

O'Brien TF, DiGiogio J, Parsonnet FC, Kass EH, Hopkins JD. Plasmid diversity in *Escherichia coli* isolated from processed poultry and poultry processors. Vet Microbiol 1993;35;243-55.

Read SC, Gyles CL, Clarke RC, Lior H, McEwen S. Prevalence of verocytotoxigenic *Escherichia coli* in ground beef, pork and chicken in southwest Ontario. Epidemiol Infect

1990;105:11-20.

Roberts TA, Britton CR, Hudson WR. The bacteriological quality of minced beef in the U.K. J Hyg 1980a;85:211-7.

Roberts TA, Macfie HJH, Hudson WR. The effect of incubation temperature and site of sampling on assessment of the numbers of bacteria on red meat carcasses at commercial abattoirs. J Hyg 1980b;85:371-80.

Smith MG, Graham A. Destruction of Escherichia coli and salmonellae on mutton carcases by treatment with hot water. Meat Sci 1978;2:119-28.

Smulders FJM. Preservation by microbial decontamination; the surface treatment of meats by organic acids. In: Gould GW, editor. New methods of food preservation. London: Blackie Academic and Professional, 1995:253-82.

Sussman M, editor. The virulence of *Escherichia coli*. London: Academic Press, 1985.

Tarkka E, Åhman H, Siitonen A. Ribotyping as an epidemiologic tool for *Escherichia coli*. Epidemiol Infect 1994;112;263-74.

United States Department of Agriculture and Food Safety and Inspection Service. National microbiological baseline data collection programme: steers and heifers, 1994.

Vorster SM, Greebe RP, Nortjë GL. Incidence of *Staphylococcus aureus* and *Escherichia coli* in ground beef, broilers and processed meats in Pretoria, South Africa. J Food Protect 1994;57:305-10.

Whelehan OP, Hudson WR, Roberts TA. Bacteriology of beef carcasses after slaughterline automation. J Hyg 1986;96;205-216.

Willshaw GA, Smith HR, Roberts D, Thirlwell J, Cheasty T, Rowe B. Examination of raw beef products for the presence of Vero cytotoxin producing *Escherichia coli*, particularly those of serogroup O157. Lett in Appl Microbiol 1993;75;420-6.

Chapter 8
Survival Of The Verotoxigenic Strain *E. Coli* O157:H7 In Laboratory-Scale Microcosms

Dr Andrew Maule

Microbial Technology Department, CAMR, Porton Down, Salisbury, Wiltshire, SP4 0JG

1 INTRODUCTION

Most strains of the bacterium *Escherichia coli* are considered to be part of the normal microbial flora of the gastro-intestinal tract of man and other warm blooded animals. However, certain strains are pathogenic and cause characteristic diarrhoeal syndromes. Those *E. coli* associated with foodborne illness are usually grouped into four categories and based on virulence properties, clinical syndrome, epidemiology and O:H serogroups. These categories are enteropathogenic *E. coli* (EPEC), enteroinvasive *E. coli* (EIEC), enterotoxigenic *E. coli* (ETEC) and enterohaemorrhagic *E. coli* (EHEC).

    E. coli O157:H7 is an EHEC. Members of this group produce toxins harmful to cultured vero cells (African green monkey kidney cells) and are thus often called verocytotoxin-producing *E. coli* or VTEC. These strains can cause haemorrhagic colitis in humans, which usually presents as bloody diarrhoea, a characteristic symptom of VTEC infection.

    E. coli O157:H7 was first isolated in 1975 (Riley *et al.*, 1983) in the USA from a woman with grossly bloody diarrhoea. Since then, VTEC, and in particular *E. coli* O157:H7 have been recognised as major foodborne pathogens, and have caused several large outbreaks of haemorrhagic colitis in the USA (Neill, 1994). It is estimated that *E. coli* O157:H7 now causes 20,000 cases of diarrhoea in the United States each year (Finelli *et al.*, 1995). In the United Kingdom the number of O157 VTEC infections have risen from a handful in the early 1980's to approximately 650 in 1994 (Anon, 1995).

    The vehicle of infection is usually undercooked food, especially minced beef products such as beefburgers (Neill, 1994). However, contaminated water has also caused a major outbreak resulting in over 240 cases of diarrhoea and four deaths (Geldreich *et al.*, 1992).

    In addition to haemorrhagic colitis, infection with *E. coli* O157:H7 can lead to other complications such as haemolytic uraemic syndrome (HUS) and thrombotic thrombocytopaenic purpura (TTP). Typically, HUS following VTEC infection presents about a week after the onset of diarrhoea, and is characterised by anaemia and renal failure. Between 2% and 7% of those infected with VTEC develop HUS, and in the UK it is a major cause of acute renal failure in children. The mortality rate of VTEC infection associated with HUS is between 5% and 10%.

    It is believed that the number of organisms required to produce infection is low, and ingestion of less than 100 organisms may be sufficient.

    There is no specific treatment for the disease, and the effect of antibiotics in modifying the course of the illness has not been established. Consequently VTEC infection represents a significant and serious public health problem. VTEC are present in the gastro-intestinal tract of normal animals, and in one study were recovered from the faeces of 8.4% of healthy dairy cows and 19% of diary heifers and calves (Wells *et al.*, 1991). Consequently cattle are a reservoir for *E. coli* O157:H7 and other VTEC and pasture land grazed by these animals may become infected with VTEC. Little is known of how VTEC may survive in soil and water ecosystems. However, *E. coli* O157:H7 survived sufficiently long in a manured garden soil to cause infection in a woman who consumed poorly washed vegetables (Cieslak *et al.*, 1993).

The potential therefore exists for the spread of VTEC from healthy animals to pasture land and then to surface waters by run-off from rainfall. The aim of this work was to investigate the survival of *E. coli* O157:H7 in various laboratory-based ecosystems representing soil, cattle faeces and slurry and river water.

2 MATERIALS AND METHODS

Soil microcosms. These comprised cores (50 mm long x 18 mm diam.) taken from a garden lawn. The cores were inserted into sterile plastic universal bottles, inoculated with 1 ml of an 18 hour culture of *E. coli* O157:H7 and incubated at 18°C under continuous illumination.

Survival of *E. coli* O157:H7 in cattle slurry. Slurry was collected from the slurry tank of a dairy farm near Salisbury, Wiltshire. A sample (100 ml) was measured into a sterile conical flask (250 ml) inoculated with 1 ml of an 18 hour culture of *E. coli* O157:H7 and then incubated in an orbital shaking incubator (100rpm) at 18°C.

Survival of *E. coli* in cattle faeces. Fresh cattle faeces were collected from a field of grazing dairy cattle on a farm near Salisbury, Wiltshire. Aliquots (10 g) of faeces were weighed into sterile plastic universal bottles, incubated with 100 µl of an 18 hour culture of *E. coli* O157:H7 and incubated at 18°C under continuous illumination.

Survival of *E. coli* in river waters. Water was collected from the River Bourne, near Idmiston, Salisbury, Wiltshire, and aliquots (500 ml) measured into sterile 2L conical flasks. These were inoculated with 1 ml of an 18 hour culture of *E. coli* O157:H7 and incubated statically at 18°C under continuous illumination.

Culture of *E. coli* O157:H7 (strain PS14). This strain was obtained from the Laboratory of Enteric Pathogens, PHLS, Colindale, London. It was cultured in aliquots of 100 ml double-strength Luria-Bertini medium in 250 ml conical flasks. This medium comprised 950 ml deionised water to which was added 20 g Bacto Tryptone (Difco), 10 g Bacto yeast extract (Difco) and 10 g NaCl, with the pH adjusted to 7.2. The flasks were incubated at 37°C in an orbital shaker (120 rpm).

Enumeration of viable numbers of *E. coli* O157:H7 in the various model ecosystems was done by inoculating dilutions onto triplicate plates of cefixime-tellurite-sorbitol MacConkey (CTSMAC) agar (Zadik, Chapman and Siddons, 1993). Dilutions were made in phosphate-buffer, manucol, antifoam (PBMA). This diluent comprised 4.5 g KH_2PO_4, 0.5 g NH_4Cl, 0.5 g NH_4SO_4, 1000 ml deionised water, 100 ml manucol (2.5% w/v aq. solution) and 1 ml GE Silicones 60 (10% w/v aq. solution), pH 7.6. For solid samples the initial 1/10 dilution was made by homogenising the sample in a Seward 400 stomacher (Seward, London).

Plates were surface-dried before use and incubated at 37°C for 24 h to allow colonies to develop. *E. coli* O157:H7 typically appeared as grey, non-sorbitol fermenting colonies with CTSMAC medium.

Table 8.1 *The numbers of viable E. coli O157:H7 surviving in laboratory-scale ecosystems after various periods of incubation*

CATTLE FAECES		CATTLE SLURRY		RIVER WATER		SOIL CORES	
No. g^{-1}	DAY	No. ml^{-1}	DAY	No. ml^{-1}	DAY	No. g^{-1}	DAY
7.1×10^7	0	1.2×10^8	0	6.7×10^6	0	8.1×10^7	0
9.7×10^7	1	1×10^9	1	4.6×10^6	2	2.2×10^8	7
2.1×10^7	6	9.8×10^8	2	4.9×10^3	6	9.3×10^7	14
2×10^7	9	7.6×10^8	3	3.8×10^3	7	3.9×10^7	27
1.5×10^7	14	1×10^4	7	2.8×10^3	9	2.8×10^7	34
5.4×10^6	20	1.6×10^3	8	5.5×10^1	13	6.0×10^6	41
2.9×10^6	27	Not detectable	9			1.1×10^7	55
5.7×10^6	34					8.7×10^6	63
$8.7x \times 10^6$	43						
3.8×10^5	54						

3 RESULTS

The recovery of viable *E. coli* O157:H7 from the four different model ecosystems is shown in Table 8.1. Each results represent the average of triplicate counts on CTSMAC agar. The minimum level of detection using this method was approximately 100 organisms/g of material. However, in the case of the river water which contained far fewer background organisms, the detection limit was approximately 20 organisms/ml.

It is evident that of all the model ecosystems tested *E. coli* O157:H7 survived best in the soil cores. In these systems the viable numbers fell from an average of 8.1 x 10^7/g to 8.7 x 10^6/g in 63 days. The organisms were less stable in cattle faeces where the average number of detectable *E. coli* O157:H7 declined from 7.1 x 10^7/g to 3.8 x 10^5/g in 54 days. In contrast to these cases *E. coli* O157:H7 seemed to show quite rapid cell death in cattle slurry and river water, where numbers declined to less than 100/g in 9 and 13 days, respectively.

4 CONCLUSIONS

VTEC and *E. coli* O157:H7 in particular is a relatively newly emerged pathogen and is the cause of haemorrhagic colitis and other more serious disease syndromes. Although the main route of infections seems to be by food, especially under-cooked beefburger there have been a number of cases where contaminated water (Swerdlow *et al.*, 1992, Moore *et al.*, 1993, Isaacson *et al.*, 1993, Brewster *et al.*, 1994) and soil-contaminated food (Cieslak *et al.*, 1993) have been involved. It is thus evident there is a need to understand more about the ecology of the organism. In particular its survival in the environment and the potential for human and animal wastes to contaminate water supplies are poorly understood.

The current study has shown that *E. coli* O157:H7 seems to survive for long periods both in cattle faeces and in soil. Thus it seems that once pasture land becomes contaminated with this organism it may remain viable for several months. During this time the organism may be ingested by other animals through the consumption of contaminated foliage, or it may be washed into surface waters by rain fall. Moore *et al.*, (1993) described the first outbreak of *E. coli* O157:H7 infection which was linked to recreational water. In this instance the disease was associated with swimming in a lake, and continued to cause infection for more than three weeks. Thus, once the water had been contaminated it is evident that the organism could survive for a sufficient length of time to cause infection. Environmental survival is particularly important in the case of *E. coli* O157:H7 because of the extremely low infective dose required to cause disease.

In addition to animal wastes, there has been one reported instance where *E. coli* O157:H7 contaminated food grown in a garden manured with contaminated cattle faeces. In this case it seems that vegetables were grown in contaminated manure-containing soil and were not adequately washed before they were eaten. This supports the finding of the present study and shows that *E. coli* O157:H7 can survive for prolonged periods in the soil and still remain infective.

When enteropathogenic microorganisms are exposed to the environment they are often injured and when attempts are made to enumerate them on selective media, as in the present study, they may die or simply not grow (Singh and McFeters, 1990). This can lead to underestimation of bacterial numbers, thus the figures given for *E. coli* O157:H7 survival in laboratory ecosystems in this study may be much lower than the real situation. In addition, Xu, *et al.*, (1982) proposed that environmental stress can lead to *E. coli* entering a state in which they are viable but will not grow on agar. Although there is apparently no work published on VTEC entering a "nonrecoverable" stage it seems possible that they too may exhibit this phenomenon. Consequently there is still much to be learned of how VTEC survive in soil and water and respond to environmental stress.

Acknowledgements

This work was supported by the UK Department of Health.

References

Anon. The Advisory Committee on the Microbiological Safety of Food. Report on verocytotoxin-producing *Escherichia coli*. London, HMSO, 1995.

Brewster DH, Brown MI, Robertson D, Houghton GL, Bimson J, Sharp JCM. An outbreak of *Escherichia coli* O157 associated with a children's paddling pool. Epidemiol Infect 1994; 112:441-7.

Cieslak PR, Barrett TJ, Griffin PM, Genstieimer KF, Beckett G, Buffington J, *et al*. *Escherichia coli* O157:H7 infection from a manured garden. Lancet 1993; 342:367.

Finelli L, Crayne E, Dalley E, Pilot K, Spitalny KC. Enhanced detection of sporadic *Echerichia coli* O157:H7 infections - New Jersey 1994. MMWR 1995; 44:417-18.

Geldreich EE, Fox KR, Goodrich JA, Rice EW, Clark RM, Swerdlow DL. Searching for a water supply connection in the Cabool, Missouri disease outbreak of *Escherichia coli* O157:H7. Wat Res 1992;26:1127-37

Isaacson M, Canter PH, Effler P, Arntzen L, Bowmans P, Heenan R. Haemorrhagic colitis epidemic in Africa. Lancet 1993;341-961.

Moore AC, Herwaldt BL, Craun GF, Calderon RL, Highsmith AK, Juranek DD. Surveillance for waterborne disease outbreaks - United States, 1991-1992. MMWR 1993;42:1-22.

Neill MA. *E. coli* O157:H7 time capsule: what did we know and when did we know it? Dairy, Food Environ Sanit 1994;14:374-77.

Padhye NV, Doyle MP. *Escherchia coli* O157:H7:Epidemiology. Pathogenesis and methods for detection in food. J Food Prot 1992;55:555-65.

Riley LW, Remis RS, Helgerson SD, McGee HB, Wells JG, Davis RJ *et al*. Haemorrhagic colitis associated with a rare *Escherichia coli* serotype. N Engl J Med 1983;308:681-5.

Singh A, McFeters GA. Injury of enteropathogenic bacteria in drinking water. In: McFeters GA, editor. Drinking water microbiology. New York: Springer-Verlag. 1990:368-79.

Swerdlow DL, Woodruff BA, Brady RC, Griffin PM, Tippen S, Donnell HD *et al.* A waterborne outbreak in Missouri of *Escherichia coli* O157:H7 associated with bloody diarrhoea and death. Ann Int Med 1992;117:812-9.

Wells JG, Shipman LD, Greene KD, Sowers EG, Green JH, Cameron DN *et al.* Isolation of *Escherichia coli* serotype O157:H7 and other Shiga-like-toxin producing *E. coli* from dairy cattle. J Clin Micrbiol 1991;29:985-9.

Xu H-S, Roberts N, Singleton FL, Attwell RW, Grimes DJ, Colwell RR. Survival and viability of nonculturable *Escherichia coli* and *Vibria cholerae* in the estuarine and marine environment. Microb Ecol 1982;8:313-23.

Zadik PM, Chapman PA, Siddons CA. Use of tellurite for the selection of verocytotoxigenic *Escherichia coli* O157. J Med Microbiol 1993;39:155-8.

Chapter 9
Faecal Coliforms In Shellfish

D.N. Lees*[1] and M. Nicholson[2]

[1] Ministry of Agriculture, Fisheries and Food, Fish Diseases Laboratory, Weymouth, Dorset DT4 9TH, UK
[2] Ministry of Agriculture, Fisheries and Food, Fisheries Laboratory, Lowestoft, Suffolk NR33 OHT, UK
© *British Crown copyright 1995*

1 INTRODUCTION

Human health problems arising from the consumption of bivalve molluscan shellfish are well recognised. In England and Wales the main commercially harvested shellfish from inshore coastal waters are : oysters (*Ostrea edulis* and *Crassostrea gigas*), mussels (*Mytilus edulis*), cockles (*Cerastoderma edule*) and clams (*Mercenaria mercenaria* and *Tapes philippinarum*). These filter-feeding bivalve molluscs cause human health problems because they concentrate and retain human pathogens derived from sewage contaminated growing waters. Unfortunately, many of the estuaries and inlets traditionally used for shellfish cultivation are also used as convenient routes for the disposal of sewage. This hazard is compounded by the traditional consumption of shellfish raw or lightly cooked. The United Nations in their report on the marine environment (1990) stated that "the present state of knowledge indicates that the most clearly identified health risk associated with coastal pollution by urban waste water is the transmission of disease by the consumption of shellfish harvested in contaminated areas" (United Nations Environment Programme, 1990).

2 HEALTH EFFECTS

Internationally, disease outbreaks can occur on an epidemic scale as graphically illustrated by an outbreak of Hepatitis A in Shanghai, China in 1988. Almost 300,000 cases were traced to the consumption of contaminated clams (Halliday *et al.*, 1991). In England and Wales outbreaks have not occurred on this scale but never-the-less continue to occur on a regular basis. Disease statistics (all illness associated with shellfish consumption) compiled by the PHLS Communicable Disease Surveillance Centre for the last decade are shown in figure 1. It should be noted that as only "outbreaks" are collated these figures probably represent a considerable under reporting. Outbreaks with a known aetiology are predominantly caused by small round structured viruses (SRSV's) of the Norwalk or Norwalk-like family which cause gastroenteritis. A smaller proportion of cases are caused by Hepatitis A virus, and only a very few cases are bacterial in origin (Figure 9.1). Although the majority of cases have an undefined aetiology, the clinical symptoms are mostly consistent with viral gastroenteritis. The available evidence therefore suggests that viral infections are the predominant cause of human disease following shellfish consumption in England and Wales.

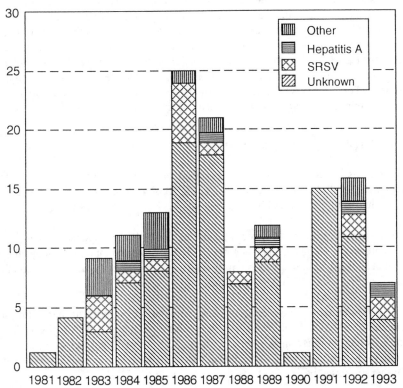

Figure 9.1 *Number and cause of outbreaks of illness associated with consumption of molluscan shellfish in England and Wales. Data compiled by PHLS Communicable Disease Surveillance Centre, Colindale.*

3 TREATMENT PROCESSES

Traditionally, three processes have been used to treat contaminated shellfish: heat treatment, depuration and relaying. Approved commercial cooking processes are required to raise shellfish meats to 90°C and to hold that temperature for 90 seconds. These heat cook parameters are very effective at inactivating pathogens and since their introduction MAFF is not aware of any infectious disease incidents associated with commercially cooked shellfish. For shellfish traditionally sold live, self-purification, either in approved tanks (depuration) or in the natural environment (relaying), can be used. For depuration, harvested animals are transferred to tanks of clean seawater where they continue to filter feed for a minimum of 42 hours during which time sewage contaminants are purged out by the normal physiological processes. During this process, seawater is irradiated by ultra-violet light to inactivate contaminants in suspension before being recirculated. Unfortunately, this process is not as effective in removing human pathogenic viruses as it is in removing faecal indicator bacteria. Numerous cases of viral gastroenteritis have been documented following consumption of depurated shellfish which meet the specified bacteriological end product standard of <230 *E.coli* or <300 faecal coliforms per 100g of shellfish flesh and fluid (see below). Relaying involves the transfer of harvested animals to cleaner estuaries or inlets where they are left for a protracted period. European Community (EC) legislation for category C areas (see below) requires that animals must be relaid for a minimum period of 2 months. A relaying period for category B shellfish is not specified in the legislation. This treatment option is little used in England and Wales principally because of the difficulties in finding under-utilised waters that are sufficiently clean and in establishing ownership rights to those waters. Category C shellfish relaid in category B waters still require depuration before they can be placed on the market.

4 STATUTORY CONTROL MEASURES

Statutory controls for molluscan shellfish are now largely determined by European legislation (Council Directive 91/492/EC) enacted in domestic legislation under the Food Safety (Live Bivalve Molluscs and Other Shellfish) Regulations 1992. The controls stipulate:-

(a) a defined end product standard for all molluscan shellfish sold to the consumer which embraces both microbiological and other criteria. The most important microbiological aspect of this is that shellfish should always comply with a standard of <230 *E.coli* or <300 faecal coliforms per 100g of shellfish flesh and fluid.

(b) the requirement to determine the degree of faecal pollution in harvesting areas according to defined faecal coliform parameters. This classification determines the level of treatment to which molluscs harvested from these areas must be subject to or, in extreme cases, prohibits harvesting. This classification of harvesting areas has far reaching consequences and is further discussed below.

(c) standards for depuration, heat-treatment and relaying.

(d) general hygiene standards for purification plants and dispatch centres.

(e) a "paper trail" to ensure that shellfish causing illness can be traceable back to the harvesting area.

(f) equivalent standards for third country imports.

In England and Wales the "Competent Authority" arrangements for this directive are a shared responsibility between Local and Port Health Authorities, the Department of Health and MAFF.

5 CLASSIFICATION OF HARVESTING AREAS

Probably the most significant new stipulation of Directive 91/492/EC is the requirement to classify harvesting areas according to sewage pollution levels. The purpose of this classification is to ensure that adequate shellfish processing occurs prior to consumption or that, in grossly polluted areas, shellfish are prevented from reaching the market for consumption. Harvesting

Table 9.1 *Summary of sampling protocol for shellfish testing for classification of harvesting areas under Directive 91/492/EC.*

Note : The Food Safety (Live Bivalve Molluscs and Other Shellfish) Regulations 1992 require the designation of bivalve mollusc production areas. This necessitates the sampling and testing of bivalve molluscs in order to determine the extent of sewage (or other faecal) contamination. This sampling only needs to be undertaken for areas from which bivalves are being taken to be placed on the market. Wild, but not farmed, scallops are excluded from these requirements. The Regulations do not necessitate the designation, and therefore classification, of areas used for the production of "other shellfish" (gastropods, etc.), although the Ministers can designate areas which from which the harvesting of "other shellfish" are prohibited.

Sample species	Different species of bivalve molluscs concentrate bacterial indicators to differing extents. Therefore, in general, each species within a production area will need to be sampled separately. Wherever possible, species should be sampled by the method normally used for commercial harvesting as this can influence the degree of contamination.
Sample sites	Samples should be taken from sentinel monitoring points agreed with the Fish Diseases Laboratory, Weymouth. Location of these sample points should take account of the location and nature of the shellfishery, the position of sewage discharges, tidal flows and local topographical details. When sampling by dredging the dredging run should include the nominal sample point, or be in the near vicinity of this.
Sample frequency	a) <u>Provisional classifications</u> . In general, for new beds or areas, 10 samples should be taken from each sentinel point over a period of 3 to 4 months. Once a provisional classification has been determined sampling frequency can be reduced to monthly in order to achieve and maintain a full classification. b) <u>Full classifications</u> . Maintenance sampling should be undertaken on a monthly basis. If particular problems occur, such as unexplained increases in the extent of contamination, then the sampling frequency may need to be increased for a period of time.
Sample composition	The following sample sizes are recommended: Oysters and clams 10 - 15 animals Mussels 15 - 30 animals Cockles 30 - 50 animals
Data collection	Food authority sample point identification (or MAFF Bed ID, where known), map co-ordinates, state of tide and method of collection (hand picked, dredged, etc.) should be recorded. Meteorological information such as wind direction and approximate strength and details of recent rainfall are also helpful. Water temperature and salinity can also prove useful and should be recorded, if feasible.
Sample transport	Use a cool box with freezer packs to keep as near to 4° C as possible. Samples should be delivered to the laboratory as soon as possible - the maximum time between collection and commencement of the test should not exceed 24 hours. **Samples should not be frozen** and freezer packs should not come into direct contact with the samples.
Sample preparation and testing	This should be undertaken in accordance with the agreed method published in the PHLS Microbiology Digest 1992, 9(2), 76 - 82. Results obtained using other methods are not currently acceptable for classification purposes. It is recommended that the examining laboratory takes part in a relevant quality assurance scheme, where available.

areas are assigned to one of three categories. Category A areas have the lowest levels of faecal pollution and are defined as either < 230 *E. coli* or < 300 faecal coliforms per 100 g shellfish flesh. Shellfish from these waters comply with the end product standard and may go for direct human consumption. Category B areas comply with standards in the range 230-4600 *E. coli* or 300-6000 faecal coliforms per 100 g shellfish flesh with a 90% compliance regime. Shellfish from this category must be depurated, heat treated or relaid until they meet category A standards. Category C areas are defined as containing shellfish with up to 60,000 faecal coliforms per 100 g flesh. Shellfish from these waters must be relaid for 2 months or heat treated by an approved process to meet category A or B standards. Shellfish exceeding the upper limit of category C are prohibited for harvesting for human consumption. Fishing from a classified area may also be controlled in an emergency through the use of a temporary prohibition order.

MAFF issue annual Classification listings on the basis of bacteriological monitoring undertaken by Local and Port Health Authorities. Monitoring programs, sampling procedures and testing methodologies are performed according to agreed protocols to ensure national consistency. These protocols are based on field studies and on experience gained during implementation of this Directive and are designed to minimise data variability from testing methods, sampling procedures and inherent variability in the marine environment. A summary of the sampling protocol is listed in Table 1. The classification listing now covers more than 70 main production areas around the coast of England and Wales. The most recent classification listing, July 95, shows that 64% of designated beds within these production areas fall into category A or B, 28% fall into category C, and 7% are designated prohibited for harvesting. A "provisional" classification is given where monitoring data, for various reasons, is incomplete but sufficient to indicate a probable trend. "Seasonal" classifications can be given where monitoring results show a deterioration only during a non-commercial harvesting period and where a sufficient period has elapsed between the poor quality period and the resumption of commercial harvesting. MAFF is currently compiling comprehensive maps of all harvesting areas which include information on the location of shellfish beds, agreed monitoring points and known sewage discharges. On the basis of this mapping information, and on data generated by the shellfish monitoring programs, precise geographical descriptions of designated harvesting areas (Classification Zones) have been defined.

6 RELATIONSHIP BETWEEN *E.COLI* COUNTS IN SHELLFISH AND IN WATER

EC Directive 91/492 stipulates bacteriological criteria for shellfish flesh but these standards are not translated into an equivalent water quality standard. Unfortunately *E.coli* (or faecal coliform) counts in shellfish do not directly correlate with *E.coli* counts in water. This is problematical where predictions of shellfish quality are required to be based on existing water quality information or where water quality criteria are sought to achieve specified shellfishery classifications. Review of the scientific literature reveals little information on the relationship between bacterial counts in water and bacterial counts in shellfish. Only a limited number of laboratory and environmental studies have been reported and certainly no consensus figure for shellfish/water concentration factors has yet emerged. This prompted the establishment of a MAFF / Local Authority field survey based on monitoring already in place for the purpose of compliance with Directive 91/492/EC. This study aimed to establish the relationship between bacterial levels in water and in shellfish and to extend this to the calculation of a bacteriological water quality standard that would allow category B, or better, status to be achieved in shellfish flesh.

Six geographically separate Local Authorities took water samples simultaneously with shellfish taken for monitoring purposes. All samples were tested by the appropriate local Public Health Laboratory Service (PHLS) laboratory as routine monitoring samples. Shellfish were tested for *E.coli* by the standard approved protocol (Table 1). Water samples were tested for *E.coli* by the standard method in use in the PHLS laboratory which was either membrane filtration (MF) or multiple tube most probable number (MPN). The survey extended over a period of 32 months and generated 602 matched results spanning 3 shellfish species (mussels), *Ostrea edulis* (native flat oysters) and *Crassostrea gigas* (Pacific oysters), 6 different main locations with 46 different sampling points, 4 different PHLS laboratories and covering all seasons of the year. This wide range of parameters was felt to be sufficient to minimise bias from any particular variable.

As a preliminary assessment of these data, Figure 9.2 shows for all sites the individual numbers of *E. coli* in shellfish (n_s) plotted against the corresponding number of *E. coli* in sea water (n_w). Superimposed is a line joining the geometric means of *E. coli* in shellfish (gm_s) to the corresponding geometric means in sea water (gm_w). To allow for zeroes in the data, numbers of *E. coli* have been increased by 1 throughout. Even at this gross level of aggregation (all sites, all species) there is some tendency for n_s to increase with n_w. This is made clearer in Figure 9.3 which shows a bubble plot (bubble size proportional to sample size) of the site specific geometric means for shellfish (gm_s) plotted against the site geometric means for seawater (gm_w) on a logarithmic scale, together with a weighted linear regression line

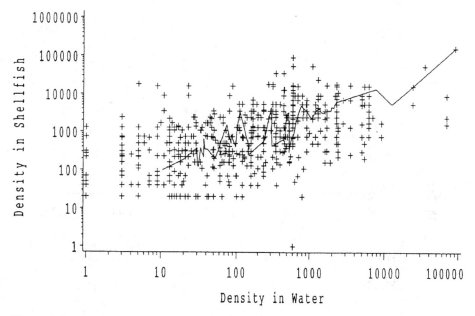

Figure 9.2 *Overall relationship between E.coli in shellfish and seawater. All individual values of E.coli in shellfish plotted against the corresponding value for seawater, with a line joining the corresponding geometric means for shellfish and for seawater at each site.*

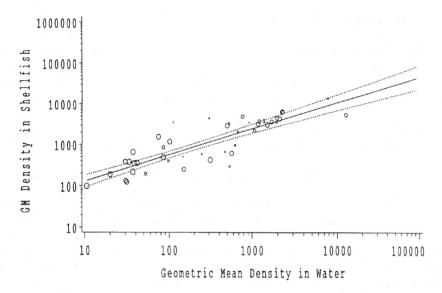

Figure 9.3 *Relationship between E.coli in shellfish and seawater at the site geometric mean level. Bubble plot (bubble size proportional to sample size) of the within-site geometric means with a superimposed regression line weighted according to sample size. Dotted lines are the 95% confidence limits of the regression line.*

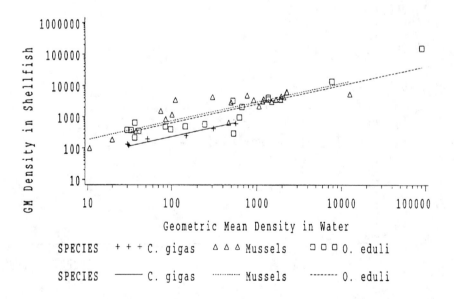

Figure 9.4 *Relationship between E.coli in shellfish and seawater by shellfish species. Site geometric means for each species plotted against the corresponding geometric mean for seawater with a superimposed weighted regression line for each species.*

Table 9.2 *Analysis of deviance*

Source	d.f.	Deviance	% Probability
E. coli in sea water	1	123.2	0.0
Residual	44	77.6	0.1
Different α's	2	24.0	0.0
Residual	42	53.6	10.8

Table 9.3 *Comparison of model fit by shellfish species*

Species	Compliance Rate for $gm_w = 100$	gm_w for Compliance Rate = 90%
C.gigas	98%	2005
O.edulis	91%	134
mussels	84%	38
All species	90%	101

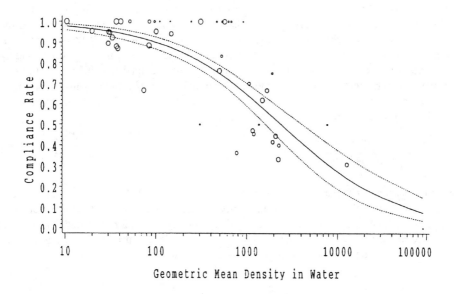

Figure 9.5 *Category B compliance rate at each site (all species). Bubble plot (bubble size proportional to sample size) of category B compliance rates of shellfish at each site plotted against corresponding geometric mean values for seawater. Superimposed is the fitted logistic regression (weighted by sample size) with its 95% confidence limits (dotted lines).*

[$\log_{10}(gm_s) = 1.51 + 0.62 \log_{10}(gm_w)$] and its 95% confidence limits. This represents the relationship <u>between</u> sites in their geometric mean levels of *E. coli* in shellfish and water, and there is a clear indication that sites with generally high levels of *E. coli* in seawater have generally high levels in shellfish, and *vice versa*. This analysis pools all shellfish species. However, scrutiny of site geometric means revealed consistent differences in the relationship according to species.   Figure 9.4 shows geometric mean levels of *E. coli* in shellfish and water for each site plotted according to species tested. Also shown are the weighted linear regression lines for each species. It was particularly noticeable that there was no significant difference between the calculated slopes for each individual species. This analysis shows that a consistent and statistically significant relationship exists between *E.coli* levels in shellfish and in water when data is aggregated at the geometric mean level. However, the actual bioaccumulation ratio will depend on the shellfish species tested and the seawater geometric mean. Figure 9.4 shows that for a seawater geometric mean of 100 on average *C.gigas* will tend to concentrate *E.coli* 2.3 times, whereas for *O.edulis* and mussels the concentration factor is 6.4 and 7.5 respectively.

7 RELATIONSHIP BETWEEN CLASS B COMPLIANCE AND E.COLI IN SEAWATER.

The above analysis establishes a relationship between counts in shellfish flesh and in seawater at the geometric mean level. However, this is not directly useful for establishing compliance with category B of the EC Directive which requires 90% of shellfish tested to comply with a limit value of 4600 *E. coli*. This was approached by modelling the proportion of shellfish values below 4600, i.e. the Class B compliance rate, which is assumed to depend on the average density of *E. coli* at a site, which we will estimate by gm_w. Let N_i be the number of animals sampled at the *i*'th station and r_i the number of animals for which the corresponding numbers of *E. coli* fall below 4600. Then, if the level in one animal is unaffected by those in another, r_i can be treated as the outcome of a Binomial trial of N_i samples with probability p of compliance and $1-p$ of non-compliance, where p depends on gm_w. The relationship between p and gm_w is assumed to be the logistic equation, i.e.

$$p = \frac{e^{\alpha+\beta\log gm_w}}{1 + e^{\alpha+\beta\log gm_w}}$$

which was found to be reasonable. Figure 5 shows a bubble plot of the observed values of r_i/N_i plotted against the geometric mean number of *E. coli* in sea water, with bubble size proportional to N_i. Superimposed is the fitted logistic regression for all of the data with its 95% confidence limits. The fitted model gives $\alpha = 5.38$ and $\beta = -0.69$. The gm_w value corresponding to a 90% compliance is obtained by solving

$$\log \frac{0.9}{0.1} = \alpha+\beta\log gm_w$$

for gm_w, giving <u>$gm_w = 101$</u>, with approximate 95% confidence limits of 63 and 162. Roughly then, we can say that to achieve a 90% compliance of Category B at a station, the average number of *E. coli* in sea water should be no greater than a geometric mean of 100, with 95% confidence limits for this estimate of approximately 60 - 160.

The model described was fitted to pooled data regardless of shellfish species. However, Figure 9.4 shows that species specific effects may be significant. This was investigated by extending analysis of deviance of the model to test whether the fitted compliance rate was the same for all species (Table 9.2). We see that although the effect of *E. coli* in sea water is highly significant, the residual from this model is also significant at the 5% level. i.e. there is rather more variation in the data than is predicted under the assumption that the number of animals complying follows a binomial distribution, suggesting that a more complex model is appropriate.

When the model is extended to allow the α terms to be different for each species, there is a significant improvement in the model, with no evidence of lack of fit. The fitted model with a different α term for each species is ; *C.gigas*, α = 6.46 ; mussels, α = 4.23 ; *O.edulis,* α = 4.94 and β = -0.56. We see that the values of α for mussels and *O.edulis* are similar, and smaller than the value for *C.gigas*. However, the differences between all three species were significant. Table 9.3 compares the compliance rates at different densities of *E. coli* in seawater for each species and for the overall fitted model. The geometric means corresponding to a 90% shellfish compliance rate are also given. We see that the 'all species' water quality criteria of a geometric mean of 100 would tend to be about right for *O.edulis*, would overestimate category B compliance for mussels and under estimate category B compliance for *C.gigas*. Table 9.3 gives calculated water geometric means for the species tested which would equate to a 90% compliance with category B.

8 IMPLICATIONS FOR DIRECTIVE 91/492/EC AND A WATER STANDARD

This study shows widely variable individual values for shellfish and water which determines poor predictive value for individual observations. However, data aggregation at the geometric mean level shows that a statistically significant relationship exists between *E.coli* levels in shellfish and in water. This was modelled to establish targets for geometric mean levels of *E.coli* in seawater compatible with category B compliance in shellfish. It should be noted that this can only be usefully applied when sufficient observations have been gathered to establish a reliable geometric mean. During the study strong species specific effects were noted suggesting that the three shellfish species studied bioaccumulate *E.coli* from water at different rates and to different concentrations. This suggests that for a given growing water quality the impact of the provisions of Council Directive 91/492/EC will be more stringent on the mussel industry than on the oyster, and in particular the *C.gigas*, industry.

The species dependant effects complicate both the generation and application of water quality criteria for shellfish classification and the comparison of shellfish standards with those, such as the FDA standards, based on water quality alone. It is possible to determine water quality compliance criteria separately for each shellfish species (Table 9.3); however, this raises obvious logistic complications. Possibly the best approach is to adopt a 'compromise' figure. An 'all species' standard of a geometric mean in sea water of 100 per 100ml would give between 84% (mussels) and 98% (*C.gigas*) compliance with category B in shellfish. It is suggested that this water quality criteria represents the best available balance between the species tested and the most appropriate compromise with the data to hand.

9 CONSUMER PROTECTION AFFORDED BY FAECAL COLIFORM MONITORING

It is appropriate to consider the degree of consumer protection afforded by monitoring shellfish for *E.coli* or faecal coliforms. In this context it should be noted that the bulk of identified disease associated with molluscan shellfish consumption is due to human enteric viral pathogens. Unfortunately, it is well documented that bacteria may behave differently to viruses in the marine environment, particularly with regard to their survival times. This may clearly also influence the ability of bacterial monitoring to predict viral contamination in bivalve shellfish. In addition, it is well documented from disease outbreaks that the absence of *E.coli* or faecal coliforms from purified shellfish does not provide a guarantee of consumer safety. Indeed, an *E.coli* failure in purified shellfish causing disease is extremely rare. These limitations have focused MAFF research in two priority areas; 1. the development of faecal pollution indicator organisms more representative of the behaviour of viral pathogens than the bacterial indicators currently in use; and 2. the development of methods for the direct detection in shellfish of the viral pathogens causing illness.

Attention is currently focused on the use of F+ bacteriophage as an alternative pollution indicator organism. Like the pathogenic viruses it has a single-stranded RNA genome, is of a similar size, and has other similar characteristics. In addition it is ubiquitous in sewage and is cheap and easy to assay. The behaviour of this potential 'virus indicator' has been explored in shellfish in the marine environment and during depuration. During depuration it is noticeable that F+ bacteriophage is removed much more slowly than *E. coli* following contamination by exposure to a crude sewage discharge (Doré and Lees, 1995). This effect has been observed in both oysters and mussels and under a variety of depuration conditions. These studies are consistent with those from disease outbreaks which also suggest that viruses behave differently to *E.coli* during shellfish depuration. In particular, viruses are not efficiently removed during such processing whereas bacteria, such as *E.coli* and the faecal coliforms, are. Consequently, the absence of *E.coli* from purified shellfish is no guarantee of either product safety or that the processing has been appropriately conducted in order to maximise any potential viral removal. Studies are in progress to determine whether F+ bacteriophage monitoring would offer any greater degree of consumer protection.

These limitations may also be overcome by developing methods for directly detecting the viral pathogens responsible for illness. Work has focused on the application of the Polymerase Chain Reaction (PCR) to detection of viruses in shellfish. It is not possible to grow the viruses causing human gastroenteritis in the laboratory and this limitation has handicapped progress for many years. Recently, the PCR genome amplification technique has offered a productive way forward. However, for meaningful determination of the viral hazard in shellfish gram quantities need to be analysed. This has required the development of extensive shellfish processing procedures prior to PCR. Our protocol has been developed using human poliovirus as a model (Lees *et al.*, 1994) and has been recently applied to hepatitis A virus. Application to polluted field samples has shown, on a limited number of samples, that the method is at least as sensitive as conventional enterovirus isolation (Lees *et al.*, 1995a). Research is continuing on the application of these procedures to the detection in shellfish of Norwalk and related small round structured viruses causing gastroenteritis. Very recent results have shown that Norwalk-like viruses can be detected in shellfish by PCR in both polluted field samples and in shellfish associated with outbreaks of human disease (Lees *et al.*, 1995b). These various developments may ultimately lead to better shellfish processing techniques for virus removal and to tests for better determination of disease hazard in shellfish sold for consumption.

The limitations of shellfish depuration for virus removal underpins the classification of shellfish harvesting areas under Directive 91/492 which aims to limit the degree of pollution of shellfish entering the processing chain. It is therefore imperative for consumer safety that monitoring regimes accurately reflect pollution status and hence the potential virus load. Although properly conducted faecal coliform monitoring programs generally provide an adequate assessment of shellfish pollution they do suffer from limitations in some respects. Sampling protocols and testing methodologies must be performed according to standard agreed procedures to avoid the introduction of bias, and they must be performed over a sufficient period of time to build a reliable picture of the shellfish pollution status. However, wide fluctuations in individual results should be anticipated and must be allowed for in result interpretation. These fluxes reflect the rapid equilibration of faecal coliform levels in shellfish with those of the surrounding water as demonstrated during shellfish depuration studies. The same studies have clearly demonstrated that enteric viruses are not subject to the same rapid fluxes. We have explored the behaviour of the F+ bacteriophage potential 'virus indicator' in this context. Initial studies have indicated that F+ bacteriophage levels in shellfish may not be subject to the same marked variability as *E. coli*. This may prove valuable for more accurate determination of faecal pollution in sites subject to intermittent pollution or where limited data is available. The use of such alternative indicators as an adjunct to faecal coliforms may help provide additional confidence in the accuracy of pollution monitoring for molluscan shellfish.

References

Council Directive of 15 July 1991 laying down the health conditions for the production and the placing on the market of live bivalve molluscs (91/492/EEC). Official Journal of the European Communities 1991; **L 268**: 1-14.

Doré, J.W., and Lees, D.N. Behaviour of *Escherichia coli* and male-specific bacteriophage in environmentally contaminated bivalve moluscs before and after depuration. Applied and Environmental Microbiology 1995; **61**: 2830-2834.

Food Safety (Live Bivalve Molluscs and Other Shellfish) Regulations 1992. Statutory Instrument No 3164. House of Commons Library 1992.

Halliday M.L, Kang L, Zhou T, Hu M, Pan Q, Fu T, Huang Y, Hu S. An epidemic of hepatitis Attributable to the ingestion of raw clams in Shanghai China. Journal of Infectious Disease 1991; **164**: 852-9.

Lees, D.N., Henshilwood, K. and Doré, W. Development of a method for detection of enteroviruses in shellfish by PCR with poliovirus as a model. Applied and Environmental Microbiology 1994; **60**: 2999-3005.

Lees, D.N., Henshilwood, K. and Butcher, S. Development of a PCR-based method for the detection of enteroviruses and hepatitis A virus in molluscan shellfish and its application to polluted field samples. Water Science and Technology 1995; **31**:457-464.

Lees, D.N., Henshilwood, K., Green, J., Gallimore, C.I., and Brown, D.W.G. Detection of Small Round Structured Viruses in Shellfish by RT-PCR. Applied and Environmental Microbiology 1995; *(in press)*.

MAFF, DoH and PHLS Working Group. Bacteriological examination of shellfish. PHLS Microbiology Digest 1992; **9**: 76-82.

United Nations Environment Programme. Joint Group of Experts on the Scientific Aspects of Marine Pollution: The State of the Marine Environment. UNEP Regional Seas Reports and Studies 1990; No. 115.

Chapter 10
Blooming *E. Coli*, What Do They Mean?

N.J. Ashbolt[1,2], M.R. Dorsch[1], P.T. Cox[2] and B. Banens[2,3]

[1]School of Civil Engineering, The University of New South Wales, Sydney, Australia NSW 2052.
[2]AWT EnSight, Sydney Water Corporation, Sydney NSW 2114.
[3]Murrey Darling Basin Commission, GPO Box 408, Canberra, Australia ACT 2601

1 INTRODUCTION

The water industry has relied on coliform indicators of drinking water quality almost since they were first proposed by Escherich (1885) over 100 years ago. The sub-set known as thermotolerant coliforms (largely comprised of *Escherichia coli*) are regarded as specific indicators of faecal contamination and are generally required to be absent in 100ml samples (Council of the European Communities, 1976; NH&MRC & ARMCANZ, 1994). Hence, the ability of *E. coli* and other coliforms to grow within biofilms in drinking water systems (Szewzyk *et al.*, 1994), in the absence of any apparent contamination event (Camper, McFeters *et al.*, 1991), may add unwarranted costs by increased sampling and disinfection to meet regulations rather than any health threat.

It is therefore of considerable interest to understand the source(s), ecology and potential pathogenicity of these "environmental" thermotolerant coliforms *in lieu* of dropping the coliform indicator concept for some other faecal indicator. It appears that very low concentrations of total organic carbon (TOC) in drinking water (0.05-12.2 mg l⁻¹) support the growth of heterotrophs, including coliforms within the nutrient concentrating biofilms (van der Kooij *et al*, 1982; Servais *et al.*, 1987). Furthermore, growth of these coliforms, particularly in disinfectant-protected surface biofilms (Yu & McFeters, 1994) can not be dismissed as of limited public health significance, given that toxigenic strains of *E. coli* grew in simulated water supplies (Camper *et al.*, 1991; Szewzyk *et al.*, 1994) and could easily come from domestic cattle or other animals (Zhao *et al.*, 1995) associated with catchment waters. On the other hand, the outbreak of *Enterobacter cloacae* in New Haven County was clearly shown to be due to a range of clinical isolates differing from the clonal strain which grew within the distribution system at the time of the outbreak (Edberg *et al.*, 1994).

Coliforms well in excess of "normal" background concentrations are, however, common in raw drinking waters from the tropics (McNeill, 1992), probably due to growth in tropical soils which may wash into waterways (Fujioka *et al.*, 1992). Nonetheless, Means (1983) reported up to 80,000 coliforms 100ml⁻¹ in the San Joaquin Reservoir which services the Orange County Water District, California and Mackay and Ridley (1983) estimated up to 296,000 cfu of *E. coli* 100ml⁻¹ in Lake Burragorang, Sydney's major raw water source.

This paper summarises ten year trends of coliforms in Lake Burragorang and presents initial molecular studies aimed at differentiating environmental isolates of *E. coli* form faecal strains.

2 MATERIALS AND METHODS

2.1 Lake Burragorang

Lake Burragorang was formed in 1961 when Warragamba Dam was completed. The lake is approximately 48km long by 0.6km wide, covers 75 km^2 and has a capacity of 2,057,000 Ml. The total catchment of about 9,000 km^2 is protected from human recreational activity, yet it receives treated sewage effluent in three of its five input streams some 50-60km upstream of the dam wall. Furthermore, numerous native and feral animals are present, principally the Eastern Grey Kangaroo, swans and other water fowl along with feral cattle, horses and pigs. The lake is classified as oligo-mesotrophic (5-12 mg l^{-1} phosphorous), with summer and winter surface temperatures of about 24° and 15°C respectively. Offtake water is unfiltered (until September, 1996), but chloraminated and accounts for about 75% of supply for Sydney's 3.8 million inhabitants.

2.2 Water Sampling

Water samples were collected from a boat on Lake Burragorang with the aid of a Niskin sampler. Water temperature was determined immediately following aseptic decantment of 250 ml of water for microbiological analysis. Samples were preserved according to Standard Methods for dissolved oxygen (DO) which was determined by Winkler titration back in the laboratory along with turbidity, pH and chloride within 6 h.

2.3 Microbiology

Lake water samples for microbiological analyses (250 ml) were collected weekly-bimonthly during 1972-1995 in sterile glass bottles and stored over ice in the dark for up to 4h. Presumptive counts for total coliforms, thermotolerant coliforms and faecal streptococci were undertaken on m-Endo agar, m-FC (prior to 1989 on pads soaked in m-FC broth) and m-enterococcus agars respectively and confirmed according to standard methods (APHA, 1989). In addition, selected thermotolerant coliforms were identified by the API20E system (API, System S.A., Montalieu, France).

Selected *Escherichia coli* isolates were grown on nutrient agar for 48 h at 37°C prior to ribotyping, genotyping or restriction fragment length polymorphism (RFLP) studies. Genomic DNA was isolated as described by Rainey *et al.* (1992) and ribotyped by the methods of Grimont and Grimont (1986) by Dr. M. Samadpour (University of Washington, Seattle). For genotyping, the entire 16S rDNA was directly sequenced following PCR amplification as described by Dorsch and Stackebrandt (1992). All nucleotide positions given in this paper refer to the *E. coli* number system (Brosius *et al.*, 1978).

3 RESULTS

3.1 Ten Year Trends

The eightieth and ninety-fifth percentiles for the twenty year (1972-1992) monthly mean estimates of thermotolerant coliforms in surface waters some 300m from Warragamba Dam are illustrated in Figure 10.1. A biomodal distribution of thermotolerant coliforms is evident, with an Austral Summer peak in December and a smaller peak during Autumn in April-May.

Faecal coliforms (cfu 100 ml⁻¹)

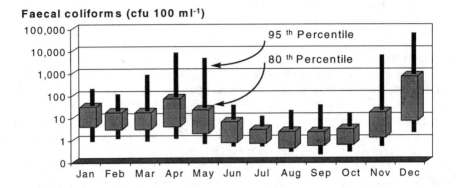

Figure 10.1 *Monthly Mean Thermotolerant Coliforms (cfu 100ml⁻¹) During 1972-1992 at Offtake of Warragamba Dam*

The upper 95th percentiles of the monthly mean rainfall during the same twenty year period also has a biomodal distribution, with peaks preceding thermotolerant coliforms by 1-2 months (Figure 10.2). Nonetheless, mean monthly rainfall is fairly constant and low (Figure 10.2).

The indirect role of rainfall prior to a bloom event is well illustrated by the 1994/95 summer bloom of *Enterobacter cloacae,* where there was practically no rainfall for the preceding five months. Nonetheless, reinstigation of lake stratification after winter overturn and consequently warmer surface waters appear to be important bloom precursors. Furthermore, the major bloom of December 1978 to January 1979 (mean counts of *E. coli* serotype O8 from 23,000 to 296,000 100 ml⁻¹) followed very heavy rainfall during March, and less heavy rainfall in July and December 1978. Associated with this rainfall were increases in water turbidity, phosphorus (from 5 to 35 mg l⁻¹), and chlorophyll a (Mackay and Ridley, 1983). The latter was primarily due to growth of the alga *Volvox* sp., with chlorophyll a increasing from a baseline of about 5 mg l⁻¹ to 32 mg l⁻¹(Cullen & Smalls, 1981).

Rainfall (mm)

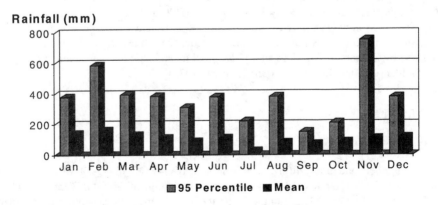

Figure 10.2 *Monthly Rainfall (mm) During 1972-1992 at Warragamba Dam*

Table 10.1 *Water Quality Parameters at Warragamba Dam, 19 December, 1994*

Depth (m)	Temperature (°C)	DO (mg l^{-1})	Turbidity (NTU)	Chloride (mg l^{-1})	pH	Coliforms[a] (cfu 100ml^{-1}) 19/12	21/12/94
3	22.8	9.6	1.3	25	8.15	6	3000
18	14.8	7.9	0.8	24	7.45	< 1	290
48	12.6	7.4	0.7	24	7.33	< 1	< 1
60	12.4	7.4	1.1	25	7.30	< 1	< 1

[a]Thermotolerant coliforms. 1000 thermotolerant coliforms at 6m on 19/12/94, hence resampled on the 21/12/94.

Blooms of thermotolerant coliforms appear to commence during stratification, when the surface (to a depth of 9-12m) water temperature rises above about 18°C. No rainfall preceded the 1994/95 bloom; however, strong SW winds did, which depressed the thermocline from the normal 9-12m to 36-40m depth at the dam wall sampling site. Water quality conditions at the commencement of the bloom (19/12/94) are given in Table 10.1.

Table 10.2 *Phenotypic and Genotypic Characterisation of Isolates*

Isolate	Source	Ribotype[a]	Genotype[b]	API 20E
ATCC 13706	ATCC	A:A1	T	1044552
Ec3	Bondi Beach sand	A:A2		1044552
Ec4	Bondi Beach sand	A:A3		1144552
Ec5	Bondi Beach sand	B:B2		5144532
Ec6	Warragamba Dam 12/91	C	T	5044552
Ec7	Warragamba Dam 12/91	B:B2	V2	5044553
Ec8	Warragamba Dam 12/91	B:B3	V2	1004552
Ec9	Warragamba Dam 12/91	B:B4	V2	5144552
Ec10	Warragamba Dam 12/91	B:B5	V2	1004552
Ec11	Warragamba Dam 12/91	B:B5	T	5044553
Ec12	Warragamba Dam 12/91	B:B5	T	5044553
Ec13	Warragamba Dam 12/91	B:B5	T	5044552
Ec14	Warragamba Dam 12/91	B:B2	V2	5044553
Ec15	Warragamba Dam 12/91	B:B7	V2	1004553
CN13	CN13 Puerto Rico[c]	A:A4		1044552
M10	M10 Puerto Rico	A:A5	T	5144572
M1M	M1M Puerto Rico	A:A6	T	5144572
M1N	M1N Puerto Rico	A:A7	V1	5144572
M1S	M1S Puerto Rico			5144572
M2M	M2M Puerto Rico		T	5144572
M2X	M2X Puerto Rico		T	5144572
M2Y	M2Y Puerto Rico		T	5144572
M1B	M1B Puerto Rico		T	5044152
M1C	M1C Puerto Rico		T	5144572

[a] Arbitrary ribotypes based on RFLP's patterns, three main groups (A-C) and variants within.
[b] Genotype based on variations in 16S rDNA as described in Table 3.
[c] Isolates from Dr. Gary Toranzos, University of Puerto Rico, Rio Piedras.

3.2 Identification of Isolates

Isolates from bathing waters were generally API 20E profile 1044552 or 5144572 and belonged to ribotype A along with the type strain (Table 2). The 16S rDNA sequence of these ribotype A isolates was the same as the type culture (T) for all, but one of the Puerto Rico isolates (variant V1, see Table 10.3). On the other hand, the Warragamba dam isolates were largely in ribotype B (one in a third group, C), but mixed with regard to the type strain genotype and variant V2.

4 DISCUSSION

Blooms of *E. coli* in Lake Burragorang in 1978/79 and 1981 occurred in the presence of other coliforms, such as *Citrobacter freundii* (API profiles 1004510, 1404572) (Mackay & Ridley, 1983) and *Enterobacter cloacae*, as well as *Aeromonas hydrophila* at up to 1,600 cfu 100ml⁻¹ (isolated on m-Endo agar, API profiles 3047166, 3047124, 3047127, 3247144 & 7247124) (L. Bowen, *pers. comm.*). Nonetheless, *E. coli* peaked at 298,000 cfu 100 ml⁻¹ throughout the top 12m of the water column and persisted for more than a month at over 5,000 cfu 100 ml⁻¹.

The L. Burragorang isolates reported by Mackay and Ridley (1983) were identified as *E. coli* based on API profiles, phage groupings (A1, A2, A4, B1, D1 or S) and by the fact that the majority were serotyped as group O8, although H10, O145, O78 were present. Serotype O8 is common in cattle faeces (Bettelheim, *pers. comm.*) and reported as a serotype of *E. coli* which grew in 40°C plus pulp and paper wastewater from Finland (Niemi, Niemelä, Mentu & Siitonen, 1987). Most of the Finish *E. coli* isolates had API 20E profiles of 5144572, the same as 70/82 of Mackay and Ridley's L. Burragorang isolates, although they also isolated strains with profiles of 1044552 and 5044542. Subsequent to the 1978/79 bloom reported by Mackay and Ridley, the December 1981 bloom of *E. coli* was dominated by the indole negative isolate (1004510) and again, an isolate with API profile 5044542 (Lance Bowen, *pers. comm.*). Hence, it was interesting that the Puerto Rico isolates were also API 5144572, whereas the lake isolates used in this study (from December, 1991) were of different profiles, but predominantly API 5044553.

From the current and previous studies (Mackay and Ridley, 1983; Niemi *et al.*, 1987) it is clear that there are strains of *E. coli* able to grow in waters from 15-42°C. Blooms of *E. coli* require organic carbon, which in the 1994/95 bloom could have, in part, come from a prior bloom of coliforms, or as in the case of the 1978/79 bloom, from green algae. Surface

Table 10.3 *Differences in the 16S rDNA of E. coli from Clinical and Environmental Isolates*

Group	16S rDNA Base Positions							
	250	273	1002	1006	1010	1019	1023	1038
Clinical[1]	**A**	**T**	**G**	**G**	**T**	**A**	**T**	**C**
V1	**A**	**T**	A	C	C	G	G	T
V2	T	A	A	C	C	G	G	T

[1]Clinical strain represented by ATCC 13706. Regions identical to the clinical strain are shown in bold.

water temperatures of over 18°C also appear important for the onset of blooms in each December.

If sewage/faeces was the source of the lake *E. coli,* then the 1978 bloom of nearly 300,000 cfu 100 ml^{-1} should have occurred in the presence of about 8,000 cfu of faecal streptococci (given the ratio of thermotolerant coliforms : faecal streptococci in sewage or stormwaters of 5-35). The absence of faecal streptococci is further justified by the realization of the quantity of faecal material required to give the densities observed. For example, given the distribution of *E. coli* in Lake Burragorang for say the 1981 bloom, some 300 tonnes of human faeces, or 12,900 tonnes of cow faeces, or 296,000 tonnes of horse, pidgeon or sparrow faeces would have to have been added (Lance Bowen, *pers. comm.*).

Work is continuing in an attempt to identify a region of the rRNA operon (from 16S through to end of 23S rRNA genes) to target PCR primers or an oligonucleotide probe to discriminate between faecal and environmental *E. coli.* Differences between these strains were identified by ribotyping and by regions of the 16S rDNA of *E. coli* not previously reported as being variable. Hence, there is reason to believe that useful differences between environmental and clinical isolates may be identified, such as in the spacer region between the 16 and 23S rDNA genes. It is considered important to develop a probe not only to rapidly discriminate between faecal and environmental isolates, but also to reduce the current underestimate of *E. coli* by culture methods, which miss injured (McFeters, 1990) as well as viable but nonculturable cells (Barcina *et al.*, 1990; Byrd *et al.*, 1991; Lewis *et al.*, 1991; Davies *et al.*, 1995).

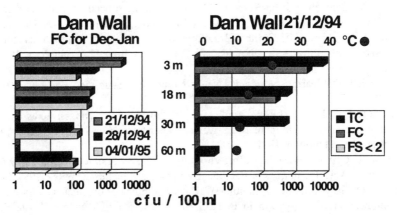

Figure 10.3 *Microbiological Water Quality and Temperature versus Depth During 1994-1995 Coliform Bloom near Warragamba Dam*

Acknowledgements

This study was partly undertaken by funding from Sydney Water Corporation (formally Water Board, Sydney, Illawarra and Blue Mountains) and a joint Australian Research Council-Industry (AWT-EnSight) grant to the first three authors. The authors would also like to thank Dr. Gary Toranzos for supply of Puerto Rican *E. coli* isolates.

References

APHA. Standard Methods for the Examination of Water and Wastewater. 17th ed. American Public Health Association, Washington D.C., 1989.

Barcina I, Gonzalez JM, Iriberri J, Egea L. Survival strategy of *E. coli* and *Enterococcus faecalis* in illuminated fresh and marine systems. Journal of Applied Bacteriology, 1990;63, 189-198.

Brosius J, Palmer JL, Kennedy JP, Noller HF. Complete nucleotide sequence of a 16S ribosomal RNA gene from *Escherichia coli*. Proceedings of the National Academy of Science U.S.A., 1978;75, 4801-4805.

Byrd JJ, Xu H-S, Colwell RR. Viable but nonculturable bacteria in drinking water. Applied and Environmental Microbiology, 1991;57, 875-878.

Camper AK, McFeters GA, Characklis WG, Jones WL. Growth kinetics of coliform bacteria under conditions relevant to drinking water distribution systems. Applied and Environmental Microbiology, 1991;57, 2233-2239.

Council of the European Communities. Council Directive of 8 December 1975 Concerning the Quality of Bathing Water (76/160/EEC). Official Journal of the European Communities, 1976;19, L31/1-L31/7.

Cullen P, Smalls I. Eutophication in semi-arid areas, in the Australian Experience. WHO Water Quality Bulletin, 1981;63, 79-83.

Davies CM, Apte SC, Peterson SM. Beta-D-galactosidase activity of viable, non-culturable coliform bacteria in marine waters. Letters in Applied Microbiology, 1995;21(in press).

Dorsch M, Stackebrandt E. Some modifications in the procedure of direct sequencing of PCR amplified 16S rDNA. Journal of Microbiological Methods, 1992;16, 271-279.

Edberg SC, Patterson JE, Smith DB. Differentiation of distribution systems, source water, and clinical coliforms by DNA analysis. Journal of Clinical Microbiology, 1994;32, 139-142.

Escherich T. Die Darmbakterien des Neugeborenen und Säuglings. Fortschr. Med., 1885;3, 515.

Fujioka R, Fujioka C, Oshiro R. Application of *Clostridium perfringens* to assess the quality of environmental and recreational waters. Water Resources Research Center, University of Hawaii, Honolulu, 1992.

Grimont F, Grimont PAD. DNA fingerprinting. In: Stackebrandt E, Goodfellow M, editors. Nucleic Acid Techniques in Bacterial Systematics. Chichester, Wiley, 1991:249-280.

Lewis GD, Grey-Young GM, Loutit MW. Bacterial indicator organisms in marine water: the implications of non-growth, In: Bell RG, Hume TM, Healy TR. editors. Coastal Engineering - Climate for Change. Proc. 10th Australasian Conf. Coastal and Ocean Engineering, Auckland, New Zealand, 2-6 December, 1991. Hamilton, Water Quality Centre Publication No. 21, DSIR, 1991:313-315.

Mackay SJ, Ridley JP. Survival and regrowth of *Escherichia coli* in Lake Burragorang, In Proc. 10th Annual Conference Australian Water and Wastewater Association, Sydney. AWWA, Sydney, 1983: 46-1 to 46-12.

McFeters GA. Enumeration, occurrence, and significance of injured indicator bacteria in drinking water, In: McFeters GA. editor. Drinking Water Microbiology. New York, Springer-Verlag, 1990:478-492.

McNeill AR. Recreational water quality, In: Connell DW, Hawker DW editors. Pollution in Tropical Aquatic systems. Boca Raton, CRC Press Inc, 1992:193-216.

Means EG. San Joaquin Reservoir Total Coliforms. Internal Report, Orange County Water District, Fountain Valley, CA, 1983.

NH&MRC, ARMCANZ. Austalian Drinking Water Guidelines. Draft June, 1994 ed. National Health and Medical Research Council, Agriculural and Resource Management Council of Australia and New Zealand, Canberra, 1994.

Niemi RM, Niemelä S, Mentu J, Siitonen A. Growth of *Escherichia coli* in a pulp and cardboard mill. Canadian Journal of Microbiology, 1987;33, 541-545.

Rainey FAR, Dorsch M, Morgan HW, Stackebrandt E. 16S rDNA analysis of *Spirochaeta thermophila*: its phylogenetic position and implications for the sytematics of the order *Spirochaetales*. Systematic and Applied Microbiology, 1992;15, 197-202.

Servais P, Billen G, Hascoet MC. Determination of the biodegradable fraction of dissolved organic matter in water. Water Research, 1987;21, 445-450.

Szewzyk U, Manz W, Amann R, Schleifer K-H, Stenström T-A. Growth and in situ detection of a pathogenic *Escherichia coli* in biofilms of a heterotrophic water-bacterium by use of 16S- and 23S-rRNA-directed fluorescent oligonucleotide probes. FEMS Microbiology Ecology, 1994;13, 169-176.

van der Kooij D, Oranje JP, Hijnen WA M. Growth of *Pseudomonas aeruginosa* in tap water in relation to utilization of substrates at concentrations of a few micrograms per liter. Applied and Environmental Microbiology, 1982;44, 1086-1095.

Yu FP, McFeters GA. Physiological responses of bacteria in biofilms to disinfection. Applied and Environmental Microbiology, 1994;60, 2462-2466.

Zhao T, Doyle MP, Shere J, Garber L. Prevalence of enterohemorrhagic *Eschericher coli* O157:H7 in a survey of dairy herds. Applied and Environmental Microbiology, 1995; 61, 1290-1293.

Natural Waters

Chapter 11
Microbiological Indicators Of Recreational Water Quality

David Kay and Mark Wyer

Centre for Research into Environment and Health, The Environment Centre, University of Leeds, Leeds, LS2 9JT, UK

1 INTRODUCTION

The use of microbiological indicators in defining standards for recreational waters is now well established. A multitude of epidemiological studies have been used to inform standards used by public health and environmental control agencies around the world. However, conducting such studies has proven logistically difficult and there has not, to date, emerged either a uniform accepted international protocol or a consistent dose-response relationship linking a single microbial indicator to disease risk in the bather population. This paper examines the epidemiological approaches taken in past attempts at the quantification of the health effects of recreational water exposure.

2 HISTORY

Table 11.1 summarises 28 epidemiological investigations into the heath effects of recreational water exposure. The majority of these have not resulted in a clear dose-response curve defining a probabilistic relationship between water quality and health. As such, they have only rarely been utilised in the design of standard systems for bathing waters.

The measurement of thermotolerant coliform organism concentration is a uniform feature of these studies for two principal reasons. First, this group is the most widely accepted and measured indicator of sewage contamination and, second, it has formed the basis of recreational water quality standards defined by control agencies in North America (NTAC, 1968), Europe (Anon, 1976) and by WHO (1972).

Several studies, notably those of Cheung *et al.* (1990) in Hong Kong, Corbett *et al.* (1993) in Australia and von Shirnding *et al.* (1993) in South Africa, have suggested a strong link between thermotolerant coliform concentrations and minor symptom reporting amongst the recreator population. However, a closer examination of Table 1 indicates both a range of microbial indicators with variable predictive power and a number of illnesses, with clearly different aetiologies, have been attributed to recreational water exposure.

Interestingly, Kueh *et al.,* (1995) found a higher correlation between turbidity in Hong Kong sea waters and gastroenteritis amongst their bathing cohort than any of the microbiological indicators enumerated. Given the fact that none of the commonly used microbial indicators measured in these studies are the aetiological agents causing the reported symptoms, there is logic in seeking a parameter, such as turbidity (i.e. nepthalotropic turbidity

Table 11.1 *Epidemiological studies linking microbiological quality of recreational waters and health impacts.*

Author	Date	Nation	Fresh/Sea	Indicator	r^2	Symptoms
Stevenson	1953	USA	both	total coliform	NR	ENT/GI/R
PHLS	1958	UK	sea	total coliform	NR	O
Cabelli	1982	USA	both	enterococci	.56	GI
Seyfried	1985	Canada	fresh	total staphylococci	.19	R/GI
				faecal coliform	.08	
				faecal streptococci	.03	
Lightfoot	1989	Canada	fresh	age	NR	R/GI
				contact person		
				interviewer		
Cheung	1990	Hong Kong	sea	E. coli	.53	GI+S
				Klebsiella	.34	GI+S
				faecal streptococci	.42	GI+S
				Enterococci	.36	GI+S
				Staphylococci	.44	ENT
				Pseudomonas aeruginosa	.38	GI+S
				Candida albicans	.30	O
				total fungi	.20	E+O
El Sharkawi	1983	Egypt	sea	enterococci	.79	GI
				E. coli	.77	
Fattal	1986	Israel	sea	enterococci	NR	GI
				E. coli		
				S. aureus		
				P. aeruginosa		
Mujeriego	1982	Spain	sea	faecal streptococci	NR	S/E/ENT/GI
Foulon	1983	France	sea	faecal streptococci	NR	E/S/GI
				total coliforms		
				faecal coliforms		
Brown	1987	UK	sea	NR	NR	GI
Ferley	1989	France	fresh	total coliforms	.21	All + S
				faecal coliforms	.45	S
				faecal streptococci	.38	GI
				Aeromonas Spp.	.26	S
				P. aeruginosa	.53	S
Balarajan	1991	UK	sea	total coliform	NR	E/GI/ENT/R
				faecal coliform		
				faecal streptococci		
Alexander	1991	UK	sea	total coliforms	NR	ENT,GI,S,O
				faecal coliforms		
				faecal streptococci		
				Salmonella		
				enterovirus		
Jones	1991	UK	sea	total coliforms	NR	E,S,ENT,R
				faecal coliforms	NR	
				faecal streptococci	NR	
Fattal	1991	Israel	sea	*S. aureus*	.62	GI
				enterococci	-.02	GI
				E. coli	-.06	GI
				total staphylococci	NR	
				P. aeruginosa	NR	
Fewtrell	1992	UK	fresh	total coliforms	NR	E,S,ENT,R
				faecal coliforms	NR	
				faecal streptococci	NR	
				total staphylococci	NR	
				P. aeruginosa	NR	
				enterovirus		

					NR	
Schirnding	1992	S Africa	sea	faecal coliform	NR	GI,R,S
				E. coli		
				faecal streptococci		
				total staphylococci		
Fewtrell	1993	UK	fresh	total coliforms	NR	E,S,ENT,R
				faecal coliforms	NR	
				faecal streptococci	NR	
				total staphylococci	NR	
				P. aeruginosa	NR	
Corbett	1993	Australia	sea	faecal coliform	NR	E,ENT,R
				faecal streptocci	NR	
Fleisher	1993	UK	sea	faecal streptococci	NR	GI
Harrington	1993	Australia	sea	faecal coliform	NR	GI,R,O
				faecal streptococci		
				C. perfringens		
Schirnding	1993	S Africa	sea	faecal coliform	NR	GI,R,S
				E. coli		
				faecal streptococci		
				total staphylococci		
Kay	1994	UK	sea	faecal streptococci	NR	GI
Asperen	1995	Netherlans	fresh	P. aeruginosa	NR	ENT
Fleisher	1996	UK	sea	faecal streptococci	NR	E,S,ENT,R
				faecal coliform		

List of Symptoms:

E   =   eye symptoms
S   =   skin complaints
GI   =   gastrointestinal symptoms
ENT =   ear nose and throat symptoms
R   =   respiratory illness
O   =   Other

Sources All named authors and Shuval (1986)

r^2 =   coefficient of determination of least squares regression models (i.e. a measure of the degree of
          explanation in the dependent variable (illness) provided by the independent variable (water quality))
NR =   not reported

units (NTU) in Kueh *et al.*, 1995), which might only indicate a general level of contamination but does provide the potential for 'real time' data acquisition which is not lagged by the significant time periods required for accurate microbial enumeration. This is important if the aim of the regulatory authority is to predict instantaneous risk to the bather and use water quality information for 'beach management' rather than retrospective 'compliance assessment.'

3 FROM EPIDEMIOLOGY TO STANDARDS

Some authorities have used the results of epidemiological investigations to derive water quality standards. An example is the work of Stevenson (1953) which underpins the National Technical Advisory Committee (NTAC) (1968) standards later adopted by USEPA. Stevenson examined the health effects of bathing at (i) two beaches near Chicago on Lake Michigan, (ii) in the Ohio River compared to a chlorinated recirculating swimming pool and (iii) tidal waters in Winchester County N.Y. at New Rochelle and Mamaroneck. The approach taken was to select two bathing areas from each water type with substantial differences in total coliform (TC) bacterial concentrations as measured by MPN techniques. The selected bathing waters were adjacent to urban centres and used by the resident population.

None of these three studies indicated a statistically significant elevation in disease reporting between the 'clean' and 'polluted' sites when the data were first examined (Kay and Wyer, 1993). However, more detailed examination of these data, in particular the effects of swimming in waters on days of high and low pollution, revealed significant increases in reported symptoms when observed water quality exceeded 2300 TC organisms per 100ml (attack rates 122 per 1000 bathers). This analysis was completed using 3 high pollution and three low pollution days for each of the two Lake Michigan beaches. Chicago North beach had a geometric mean concentration of TC organisms of 730 per 100ml during the three polluted days and only 31 per 100ml on the clean days. This was associated with attack rates of 99 per 1000 for the subsequent three day period for bathers at the polluted beach and 87 per 1000 for users of the unpolluted site. At the Chicago South beach corresponding figures for the three clean and three polluted days of observation were as follows; (i) polluted geometric mean TC organisms = 2,300 per 100ml and three day disease attack = 121 per 1000 (ii) clean geometric mean TC organisms = 43 per 100ml and three day disease attack = 85 per 1000. Parallel analyses of clean and polluted days were completed for the Ohio river and the tidal sites but no significant differences in illness attack rates were observed for either site which could be explained in terms of water quality differences. The only other significant finding of this early study related to the expected gastroenteritis rate amongst river swimmers who reported 32% more gastroenteritis than would be expected given their demographic composition. The tentative conclusion was therefore drawn that excess gastroenteritis might occur in swimmers exposed to waters with   median total coliform densities over 2,700 per 100ml. Commenting on this finding Stevenson (page 537) noted that;

> "*the result must be treated with great caution because of the small number of cases involved. It roughly corresponds to a probability of 5% that such a result could occur as an effect of chance.*"

This work was the first prospective epidemiological study into bathing related illness and was able to demonstrate a water quality-health effect only by extreme selection of data for the Lake Michigan and Ohio River sites. No health effect could be identified for the marine waters examined.

The Stevenson (1953) study was used as the foundation for the first standard systems suggested by the NTAC. The Ohio river median value of 2,700 TC organisms per 100ml was chosen as the starting point for this standard system. Using the total coliform to faecal coliform ratio for Ohio river site, the median value of 2,300 to 2,700 total coliforms was converted to a faecal coliform concentration of 400 per 100ml. This was assumed to be the level of pollution at which health effects were detected. A safety margin was applied by defining the maximum geometric mean value of the concentration of faecal coliform organisms at 200 per 100ml (i.e. 400 divided by 2). It was further felt that recreational waters should not present a health risk more than 10% of the time. As such, the following standard was adopted;

> "*Faecal coliforms should be used as the indicator organism for evaluating the microbiological suitability of recreation waters. As determined by multiple tube fermentation or membrane filter procedures and based on a minimum of not less than five samples taken over not more than a 30-day period, the faecal coliform*

content of primary contact recreation waters should not exceed a
log mean of 200 / 100ml, nor shall more than 10% of total samples
during any 30-day period exceed 400 / 100ml."
(USEPA, 1986:2)

The extent to which the epidemiological foundation provided a sufficiently firm basis
for these standards was subsequently questioned by Moore (1975) and Cabelli *et al.* (1982).
In effect, the NTAC (1968) standards are based on the identification of a threshold or 'effect'
level at 2,300-2,700 total coliforms per 100ml. Little is known of the magnitude of the effect
above these levels (i.e. the shape of the dose-response curve) which could give an indication
of 'disease burden' for a range of poorer water quality exposures. If the aim is to characterise
the risk experienced by a population using a beach over a defined time period, it is clearly
preferable to base standards on a dose-response relationship.

4 DOSE-RESPONSE MODELS

Dose-response relationships between any environmental pollution and defined health effects
from drinking water, recreational water or, indeed, air pollution are very rare. It is more
common in defining standards for regulatory authorities to determine an 'effect' level, using
an animal model or epidemiological study, then choose a safety margin and define a standard
accordingly. However, this approach does not facilitate the quantification of population 'risk'
or, more usefully, 'disease burden' attributable to recreational water exposures.

Dose-response models linking quantified disease burden with water quality have been
reported by several authors including Cabelli *et al.*, (1982, 1983), Dufour (1984), Cheung *et
al.*, (1990) and Kueh *et al.*, (1995) who quantified 'swimming associated' gastroenteritis (GI)
from indicator organism concentration, i.e. the attack rate difference between bather and non-
bather groups. Many methodological and statistical difficulties with this approach have been
noted by Fleisher (1990a,b, 1991, 1992). Perhaps the most significant is the difficulty in
defining the 'exposure' measure, i.e. locating the data points on the 'x' axis. For example,
Figure 11.1 is adapted from the work of Cheung *et al.* (1990). The data points represent

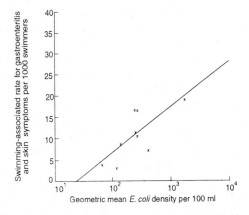

Figure 11.1 *The least squares bivariate regression relationship between geometric mean
E. coli density and the swimming associated rate for gastroenteritis (i.e. a
rate difference) (adapted from Cheung et al., 1990)*

seasonal geometric mean *E. coli* concentrations in bathing water for nine bathing locations, plotted against the combined symptom output of GI+Skin as a rate difference between bather and non-bather groups. Given, (i) the low number of data points and, (ii) the wide spread of data about the GM value (and hence imprecision of locating data points on the 'x' axis) it could be argued that these data contravene the requirements of the least squares bivariate regression model. It is certainly true that this analysis was completed on insufficient data items to allow appropriate statistical checks on the data's compliance with the assumptions of the least squares regression model (e.g. normality and constant variance in the residual term).

Seyfreid (1985a,b), Lightfoot (1989) and Corbett *et al.*, (1993) have applied logistic regression to predict the odds ratio of symptoms in the bather group. Data were acquired in a prospective manner, broadly following the protocol established by Cabelli et al., (1982), i.e. beachgoers were identified at the time of exposure when water quality measurements took place and subsequent telephone enquiry was used to acquire symptom data. The advantage with this statistical treatment of the data is that it overcomes some of the problems with 'rate difference' estimates of illness (i.e. the effects of variation in the 'background' or 'non-bather' rate) and, more importantly, it facilitates the quantification of the importance of factors which may confound any relationship between water quality and health. Lightfoot (1989) was the first to report a thorough analysis on the effect of 'confounders' on the relationship between water quality at freshwater locations and health impact. She observed that;

> *"there was no evidence to suggest that bacterial count contributed to the prediction of illness in swimmers. Instead, age, contact person, and interviewer, most frequently tended to be important."*
> (Page iv) and later

> *There is little evidence from the present study to support the belief that the bacterial water quality indicators investigated herein index the short-term risk of becoming ill from swimming"*
> (Page 208)

In a study of Sydney beachgoers Corbett *et al.* (1993) identified a statistically significant elevation in symptom reporting which was best predicted by faecal coliform concentration. The geometric mean concentration of faecal coliform on the day of exposure was used initially, but the arithmetic mean was reported to produce a better model fit. Figure 11.2 is adapted from Corbett *et al.* (1993) and shows the relationship between faecal coliform and the odds ratio for any health problem reported and geometric mean faecal coliform concentration for each exposure day. The model is adjusted for age, sex and swimming duration.

Whilst this type of analysis represents a clear advance over the least squares bivariate regression approach, it still has many of the drawbacks reported by Godfree *et al.* (1990); Fleisher (1990a,b, 1991, 1992) and Fleisher *et al.* (1993). The most significant is the problem of adequately representing water quality experienced by each bather. Given the temporal and spatial variability in indicator concentrations at most beaches, a daily mean value used as the 'exposure' measure would certainly present an imprecise estimate of exposure for many individuals. Corbett *et al.* (1993) attempted to overcome this point by additional water quality sampling and acquisition of data on the time of bather exposure. They also tested the use of daily maximum or arithmetic mean values as predictor variables. However, the problem remains with even these predictors since neither attributes an accurate exposure level to the individual bather unless there is no variation in microbiological water quality through the day, which is

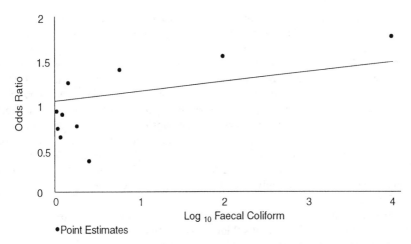

Note: ORs are adjusted for age, sex, and swimming duration; swimmers in water with 10 colony-forming units of faecal coliforms per 100 ml were the reference category.

Figure 11.2 *Logistic regression model predicting the odds ratio of a bather reporting any symptom and geometric mean faecal coliform concentration. (adapted from Corbett et al., 1993)*

very unlikely.

Given these problems, there is imprecision in the reported epidemiological dose-response relationships which make the use of these models difficult for disease burden estimation.

In a series of studies, sponsored by the UK Government between 1989 and 1992, attempts were made to overcome some of these problems using a randomised design, similar to a pharmaceutical trial (Kay *et al.*, 1994). This was completed as part of a two pronged approach involving a modified and improved application of approach pioneered by Cabelli *et al.* (1992) and the randomised design operating in parallel (Kay and Wyer, 1992; Balarajan *et al.* 1992; Fleisher, 1993; Pike, 1994). Results of the former study have yet to be fully analysed (Kay, 1994) but initial findings have been reported in Pike (1994). Results from the randomised trial were reported in Kay *et al.* (1994) and Fleisher *et al.* (1996). Both elements of the UK studies have been approved by WHO for international application (WHO, 1994).

The 'randomised trial' involved recruitment of healthy adult volunteers at seaside towns adjacent to beaches which had traditionally passed EC *Imperative* standards (Anon 1976). After initial interviews and medical checks, volunteers reported to the specified bathing location on the trial day where they were randomised into bather and non-bather groups. Bathers entered the water at specified locations where intensive water quality monitoring was taking place under the supervision of a marshal who recorded their activities. All bathers immersed their heads on three occasions. On exiting the water bathers were asked if they had swallowed water. A control group of non-bathers came to the beach and had a picnic of identical type to that provided for all volunteers. One week after exposure all volunteers returned for further interviews and medical examinations and they completed a final postal questionnaire at three weeks after exposure. Detailed water quality measurements were completed at marked locations which defined 'swim zones'. Samples were collected synchronously at locations 20m apart every 30 minutes and at three depths (i.e. surf zone, one metre and at chest depth - 1.3-1.4m). Five bacterial indicators were enumerated namely, total coliform, faecal coliform, faecal streptococci, total staphylococci and*Pseudomonas aeruginosa*. The locations and times of exposure were known for each bather and, thus, a precise estimate of 'exposure' (i.e. bacterial concentration) could be assigned to each bather.

This method has certain disadvantages over the beach survey protocol. First it was restricted to adults (i.e. over 18 years) who can give informed consent to take part in a healthy volunteer experiments which require ethical clearance from an appropriate medical research ethics committee. Second, it was restricted to beaches which traditionally pass the water quality standards in force. The former constraint is perhaps the most significant since some workers (e.g. Alexander *et al.*, 1991) have suggested that young children might be more susceptible to the types of minor infection contracted from exposure to sea water. The latter constraint has not proven a significant drawback since clear dose-response relationships have been found at beaches which currently pass Directive 76/160/EEC *Imperative* (Anon, 1976) coliform standards.

Initial examination of the data from the randomised trial experiment centred on the links between water quality and gastroenteritis (Fleisher *et al.*, 1993). The data were analysed for any relationships between water quality, as indexed by any of the five bacterial indicators measured at any of the three depths (i.e. 15 potential predictor variables), and gastroenteritis. Only faecal streptococci, measured at chest depth, provided a statistically significant relationship between water quality and the risk of gastroenteritis. This result was replicated at three of the four sites examined and at the fourth site concentrations of faecal streptococci were generally below the threshold level at which an effect was observed at the other three locations. A full statistical analysis of these data has now been reported (Kay *et al.*, 1994) which demonstrates this relationship between faecal streptococci concentrations in recreational waters and the excess probability of gastroenteritis in the exposed population (Figure 11.3). Fleisher *et al.* (1996) have extended this analysis to non-enteric symptoms which are predicted by both faecal coliforms and faecal streptococci. These reports also identified other exposure

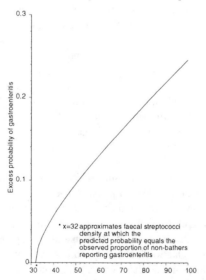

Figure 11.3 *Dose-response realtionship linking faecal streptococci exposure and the excess probability of a bather aquiring gastroenteritis. (Adapted from Kay et al., 1994)*

risks which predicted illness but which were independent of (and did not confound) the water quality - health outcome relationship. These offer useful potential markers for risks commonly encountered and accepted by beachgoing recreators in the UK.

6 CONCLUSION

There is currently (Sept. 1995) considerable debate on recreational water standards in both the USA and Europe (Kay and Wyer, 1992; Vaughan, 1993; Saliba, 1993). The epidemiological data to underpin this debate is certainly imperfect but there is a significant body of data and several dose-response relationships available in the published literature.

The recent publication of a proposal to amend the EU standards in Directive 76/160/EEC (Anon 1994a) initiated a significant review by a UK Parliamentary Select Committee which has served to collate and focus the arguments in two recent reports (Anon 1994b, 1995, Kay, 1994). The extent to which the existing epidemiological information can be integrated into this debate and any subsequent standards revision has yet to become evident. However, significant issues which should be addressed by the research community are;

(i) the lack of good dose-response relationships for fresh waters;

(ii) the absence of data on 'special interest groups' who might not be classed as bathers but who experience significant exposures to recreational waters through, for example, surfing, sail boarding, canoeing or even relatively 'low contact' activities such as sailing;

(iii) in a European context, the absence of epidemiological data from Mediterranean bathing locations;

(iv) the dose-response models provide the scope for estimating disease burden as a measure of risk, but there is little information on what constitutes 'acceptable' risk;

(v) disease burden estimates present the potential for the assessment of 'health gain' attributable to different scenarios of water quality improvement and, thus, provide for the quantification of the 'benefits' attributable to environmental improvement which can be set against the economic 'costs' which are generally the only quantified element in the decision process on infra structure expenditures.

References

Alexander, L.M. and Heaven, A. (1991) *Health risks associated with exposure to seawater contaminated with sewage: the Blackpool beach survey 1990.* Published by Lancaster University, Environmental Epidemiology Research Unit. 67p.

Anon (1976) Council Directive of 8 December 1975 concerning the quality of bathing waters. Official Journal of the European Communities No L 31 5.2.1976, p1-7.

Anon (1994a) Proposal for a Council Directive concerning the Quality of Bathing Water. *Official Journal of the European Communities No C112, 22.4.94*, p3-10.

Anon (1994b) Select Committee on the European Communities, Bathing Water. House of Lords Session 1994-5, 1st Report with evidence 6th December. HMSO, London, 149p.

Anon (1995) Parliamentary Debates, House of Lords Official Report. Hansard 564(90), 18th May, 684-708.

Anon (1995) Select Committee on the European Communities, Bathing Water Revisited. House of Lords Session 1994-5, 7th Report with evidence 21st March. HMSO, London, 32p.

Asperen van , I.A., de Rover, C.M., Colle, C., Schijven, J.F., Bambang Oetomo, S., Schellekens, J.E.P., van Leeuwen, W.J. Havelaar, A.H., Komhout, D. and Sprenger, M.J.W. (1995) *Een zwemwatergerelateerde epidemie van otitis externa in de zomer van 1994.* RIVM. Bilthoven, 133p.

Balarajan, R., Soni Raleigh, V., Yuen, P., Wheeler, D., Machin, D. and Cartwright, R. (1991) Health risks associated with bathing in sea water. *British Medical Journal 303,* 1444-1445. (for comment on this paper see Hall, A.J. and Rodrigues, L.C. (1992) Health risks associated with bathing in sea water. *British Medical Journal 304,* 572).

Brown, J.M., Campbell, E.A., Rickards, A.D. and Wheeler, D. (1987) *The public health implications of sewage pollution of bathing water.* Published by the University of Surrey, Robens Institute, Guildford, Surrey UK.

Cabelli V J, Dufour A P, McCabe L J, Levin M A. (1982) Swimming associated gastroenteritis and water quality. *American Journal of Epidemiology 115,* 606-616.

Cabelli, V.J., Dufour, A.P., McCabe, L.J. and Levin, M.A. (1983) A marine recreational water quality criterion consistent with indicator concepts and risk analysis. *Journal of the Water Pollution Control Federation 55,* 1306-1314.

Cabelli, V.J., Levin, M.A., Dufour, A.P. and McCabe, L.J. (1975) *The development of criteria for recreational waters.* In Gameson, A.L.H. (Ed.) *Discharge of sewage from sea outfalls.* Pergamon. 63-74.

Cheung, W.H.S., Kleevens, J.W.L., Chang, K.C.K. and Hung, R.P.S. (1988) Health effects of beach water pollution in Hong Kong. *Proceedings of the Institution of Water and Environmental Management.* Proceedings of the Annual Conference 376-383. See also

Cheung, Chang, K.C.K. and Hung, R.P.S. (1990) Health effects of beach water pollution in Hong Kong. *Epidemiology and Infection 105,* 139-162. and

Cheung, W.H.S., Hung, R.P.S. Chang, K.C.K. and Kleevens, J.W.L., (1990) Epidemiological study of beach water pollution and health-related bathing water standards in Hong Kong. *Water Science and Technology 23(1-3),* 243-252.

Corbett, S.J., Rubin, G.L., Curry, G.K. and Kleinbaum, D.G. (1993) The health effects of swimming at Sydney beaches. *American journal of Public Health 83,* 1701-1706.

Demarquilly, C., Boniface, B.,Maier, E.A. (1995) *Sea Water Microbiology performance of methods for the microbiological examination of bathing water Part II. Statistical Analysis.* Report EUR 16613EN. DGXII. Published by the Office for Official Publications of the European Communities, Luxembourg, 47pp, plus appendices.

Dufour, A. P. (1984) Bacterial Indicators of recreational water quality. *Canadian Journal of Public Health 75,* 49-56.

El Sharkawi, F. and Hassan, M.N.E.R. (1982) The relation between the state of pollution in Alexandria swimming beaches and the occurrence of typhoid among bathers. *Bull High Inst. pub. Hlth Alexandria 12,* 337-351.

Fattal, B., Peleg-Olevsky, E. and Cabelli, V.J. (1991) Bathers as   possible sources of contamination for swimming associated illness at marine bathing beaches. *International Journal of Environmental Health Research 1,* 204-214.

Fattal, B., Peleg-Olevsky, T. Agurshy and Shuval, H.I. (1986) The association between sea-water pollution as measured by bacterial indicators and morbidity of bathers at Mediterranean beaches in Israel. *Chemosphere 16(2-3),* 565-570.

Ferley, J.P., Zimrou, D., Balducci, F., Baleux, B., Fera, P., Larbaigt, G.,Jacq, E., Mossonnier, B.,Blineau, A. and Boudot. J. (1989) Epidemiological significance of microbiological pollution criteria for river recreational waters. *International Journal of Epidemiology 18(1),* 198-205.

Fewtrell, L., Godfree, A.F., Jones, F., Kay, D., Salmon, R.L. and Wyer, M.D. (1992) Health effects of white-water canoeing. *The Lancet 339,* 1587-1589.

Fewtrell, L., Kay, D., Newman, G., Salmon, R.L. and Wyer, M.D. (1993) *Results of epidemiological pilot studies.* In Kay, D. and Hanbury, R. (Eds.) (1993) *Recreational Water Quality Management. Vol II Fesh Water.* Ellis Horwood, Chichester. pp75-108.

Fleisher, J.M. (1990) The effects of measurement error on previously reported mathematical relationships between indicator organism density and swimming associated illness: a quantitative estimate of the resulting bias. *International Journal of Epidemiology 19(4),* 1100-1106.

Fleisher, J. M. (1991) A re-analysis of data supporting the US Federal bacteriological water quality criteria governing marine recreational waters. *Journal of the Water Pollution Control Federation* **63**, 259-264.

Fleisher, J., Jones, F., Kay, D. and Morano, R. (1993) Setting recreational water quality criteria. In Kay, D. and Hanbury (Eds.) *Recreational Water Quality Management: Fresh Water (Vol II)* Ellis Horwood. p123-136.

Fleisher, J., Jones, F., Kay, D., Stanwell-Smith, R.,Wyer, M.D. and Morano, R. (1993) Water and non-water related risk factors for gastroenteritis among bathers exposed to sewage contaminated marine waters. *International Journal of Epidemiology 22,* 698-708.

Fleisher, J., Kay, D., Jones, F.,Salmon, R.L., Wyer, M.D. and Godfree A.F.(1996) Non-enteric illnesses associated with bathing in marine waters contaminated with domestic sewage. *American journal of Public Health (in press).*

Fleisher, J.M. (1990) Conducting recreational water quality surveys. Some 47 problems and suggested remedies. *Marine Pollution Bulletin 21(2),* 562-567.

Fleisher, J.M. (1992) US Federal bacteriological water quality standards: a re-analysis of the data on which they are based. In Kay, D. (Ed.) *Recreational Water Quality Management: Coastal Bathing Waters (Vol I)* Ellis Horwood. p113-128.

Foulon, G., Maurin, J., Quoi, N.N. and Martin-Bouyer, G. (1983) Etude de la morbidite humaine en relation avec la pollution bacteriologique des eaux de baignade en mer. *Revue francaise des Sciences de L'eau 2(2),* 127-143.

Godfree, A., Jones, F. and Kay, D. (1990) Recreational water quality - the management of environmental health risks associated with sewage discharges. *Marine Pollution Bulletin* **21**, 414-422.

Harrington, J.F., Wilcox, D.N., Giles, P.S., Ashbolt, N.J., Evans, J.C. and Kirton, H.C. (1993) The health of sydney surfers: an epidemiological study. *Water Science and Technology 27(3-4),* 175-182.

Hernandez, J.F.,Delattre, J.M., Maier, E.A. (1995) *Sea Water Microbiology performance of methods for the microbiologcal examination of bathing water Part I.* Report EUR 16601EN. DGXII. Published by the Office for Official Publications of the European Communities, Luxembourg, 61pp, plus figures and tables.

Jones, F., Kay, D., Stanwell-Smith, R. and Wyer, M.D. (1991) Results if the first pilot scale controlled cohort epidemiological investigation into the possible health effects of bathing in seawater at Langland Bay, Swansea. *Journal of the Institution of Water and Environmental Management 5(1),* 91-98.

Kay, D. (1994) Summary of epidemiological evidence from the UK sea bathing research programme. Appendix 5 in Anon (1994) Select Committee on the European Communities, Bathing Water. House of Lords Session 1994-5, 1st Report with evidence 6th December. HMSO, London, p32-37.

Kay, D. and Wyer, M.D. (1992b) Recent epidemiological research leading to standards for recreational waters. In Kay. D. (Ed.) *Recreational Water Quality Management: CoastalWater (Vol I)* Ellis Horwood. p129-156.

Kay, D.,Fleisher, J.M., Salmon, R.L., Jones, F., Wyer, M.D., Godfree, A.F., Zelenauch-Jacquotte, Z. and Shore, R. (1994) Predicting the likelihood of gastroenteritis from sea bathing : results from a randomised exposure. *The Lancet 344*, 905-909.

Kueh, C. S.W., Tam, T.Y., Lee, T., Wong, S.L., Lloyd, O.L., Yu, I.T.S., Wong, T.W., Tam, J.S. and Bassett, D.C.J. (1995) Epidemiological study of swimming associated illnesses relating to bathing beach water quality. *Water Science and Technology 31(5-6)*, 1-4.

Lightfoot, N.E. (1989) *A prospective study of swimming related illness at six freshwater beaches in Southern Ontario.* Unpublished PhD Thesis. 275p.

Moore, B. (1975) *The case against microbial standards for bathing beaches.* In Gameson, A.L.H. (Ed.) *Discharge of sewage from sea outfalls.* Pergamon Press, London.

Mujeriego, R., Bravo, J.M. and Feliu, M.T. (1982) Recreation in coastal waters public health implications. *Vier Journee Etud. Pollutions, Cannes, Centre Internationale d'Exploration Scientifique de la Mer.* pp. 585-594.

NTAC National Technical Advisory Committee. (1968) *Water quality criteria.* Federal Water Pollution Control Administration, Department of the Interior, Washington DC.

P.H.L.S. (1959) PUBLIC HEALTH LABORATORY SERVICE. Sewage contamination of coastal bathing waters in England and Wales: a bacteriological and epidemiological study. *Journal of Hygiene, Cambs. 57(4)*, 435-472.

Pike, E.B. (1994) *Health effects of sea bathing (WMI 9021)-* Phase III final report to the Department of the Environment. Report No DoE 3412/2. Water Research Centre. Medmenham, UK, 138pp.

Saliba, L. (1993) Legal and economic implications of developing criteria and standards. In Kay, D. and Hanbury (Eds.) *Recreational Water Quality Management: Fresh Water (Vol II)* Ellis Horwood. p57-74.

Schirnding von , Y.E.R., Strauss, N., Robertson, P., Kfir, R., Fattal, B., Mathee, A., Franck, M. and Cabelli, V.J. (1993) Bather morbidity from recreational exposure to sea water. *Water Science and Technology 27(3-4)*, 183-186.

Schirnding, Y.E.R., Straus, N., Robertson, P., Kfir, R., Fattal, B., Mathee, A., Franck, M. and Cabelli, V.J. (1993) Bather morbidity from recreational exposure to sea water. *Water Science and Technology 27(3-4)*, 183-186.

Seyfried, P.L., Tobin, R., Brown, N.E. and Ness, P.F. (1985a,b) A prospective study of swimming related illness. I Swimming associated health risk. *American Journal of Public Health 75(9)*, 1068-1070 and 1071-1075.

Shuval, H. I. (1986) *Thalossogenic Diseases.* UNEP Regional Seas, Reports and Studies No. 79. UNEP, Athens.

Stevenson, A.H. (1953) Studies of bathing water quality and health. *American Journal of Public Health 43*, 529-538.

USEPA. (1986) UNITED STATES ENVIRONMENTAL PROTECTION AGENCY. *Ambient water quality criteria for bacteria - 1986.* EPA440/5-84-002. Office of Water Regulations and Standards Division. Washington DC 20460. 18pp.

Vaughan, J. (1993) Recreational Water Quality Standards. In Kay, D. and Hanbury (Eds.) *Recreational Water Quality Management: Fresh Water (Vol II)* Ellis Horwood. p45-54.

WHO (1994) *Microbiological quality of coastal recreational waters: Report on a joint WHO/ UNEP meeting, Athens, Greece, 9-12 June 1993,* published by WHO Regional Office for Europe with UNEP EUR/ICP/CEH 041(2).

WHO World Health Organisation (1972) *Health Citeria for the Quality of Recreational Waters with Special Reference to coastal Waters and Beaches,* 26p, WHO, Copenhagen.

Wyer, M.D. and Kay, D. (1995) New standards for Jersey bathing waters. Centre for Research into Environment and Health. Leeds University. UK 46pp.

Chapter 12
Past, Present And Future Implementation Of The EC Bathing Waters Directive In Relation To Coliforms And *E. Coli*

M.J. Figueras, F. Polo, I. Inza, N. Borrell & J. Guarro

Unit of Microbiology, Faculty of Medicine, University Rovira i Virgili, 43201 Reus, Spain.

1 INTRODUCTION

The EC Bathing Waters Directive has a common objective; for all the countries to accomplish standards of microbiological quality in order to protect the environment and public health by reducing the pollution of bathing water and by protecting it against further pollution (EEC Directive, 1976; Gilles, 1992). For almost twenty years, the EC Directive has remained as it was postulated and for some years the need for the introduction of changes that could correct some of the evident contradictions of the Directive has.been questioned. Furthermore, there is a need to evaluate whether the microbiological indicators and their numerical standards as they were postulated are the appropriate ones. This is of course the most controversial aspect, because a general agreement among the many different researchers about what could be the most adequate indicators does not exist (Saliba & Helmer, 1990; Fewtrell & Jones,1992).

The general idea that the standards are in need of updating on the basis of modern epidemiology and science exists but it is recognized that this will be a slow process (Grantham, 1992). Recently a proposal of a New Directive has been published (EEC 94/C 112/03) that has introduced some important changes in relation to the microbiological parameters. The purpose of this report is to point out the existing limitations and contradictions based on our experience and on the existing literature of the actual Directive and compare them with the new proposed modifications. Furthermore, we will also comment on the relation of *Salmonella* and *Aeromonas* with faecal coliforms.

2 BATHING WATERS DIRECTIVE LIMITATIONS AND CONTRADICTIONS

2.1 Microbiological Parameters

The existing EC Bathing Waters Directive (EEC76/160) considers the total coliforms and faecal coliforms the principal microbiological parameters.

The term total coliforms (TC) is applied to those Enterobacteriaceae belonging to the genera *Escherichia*, *Citrobacter*, *Enterobacter* and *Klebsiella* (USEPA, 1983). However, they are considered as being of poor sanitary value because they include species that are not strictly faecal (UNEP/WHO, 1983); consequently it has been suggested to eliminate the TC from the Directive (National Rivers Authority, 1990). Furthermore, it is also considered as a redundant parameter if faecal coliforms are evaluated.

The faecal coliforms (FC) are a more accepted parameter; they include thermotolerant *Klebsiella* as well as *E. coli* (Cabelli, 1983) and are considered as an indicator of recent faecal pollution. *E. coli* is the species consistently and exclusively found in faeces of warm blooded animals (Geldreich, 1970; Dufour, 1977; Geldreich, 1978; Dufour *et al.* 1981). According to the Standard Methods the faecal coliform group includes all those coliforms able to grow at 44.5°C (APHA, 1992). However, several authors synonymyzed faecal coliforms with *E. coli* even when referring to the EC Directive (Jones *et al.*, 1990; Philipp, 1991; Kay & Wyer, 1992; Rees, 1993). It is true that *E. coli* is one of the predominant species of the faecal coliform group, but not the only one that follows their characteristics according to the Standard Methods (Figueras *et al.*, 1994). This is an important point that was needed to be clarified in the Directive, because if not the results of the different EC member states could not be compared.

The existing Directive includes other microbiological parameters: faecal streptococci, *Salmonella* and enteroviruses. However, the frequency of their analysis as stated in the Directive present certain contradictions, since the measurement of these parameters is only required when the competent authority has reasons to believe that they are present or when water quality has deteriorated.

Faecal streptococci (FS) are considered by many authors as a good indicator due to the fact that they are more resistant than coliforms to environmental stress and they survive longer and so they may be more effective indicators of sewage pollution (Philipp, 1991; Rees 1993; Sinton *et al.*,1993 a,b). The FS group includes, according to the Standard Methods (APHA,1992), several species of the genus *Streptococcus*, such as *S. faecalis, S.faecium, S. gallinarum, S. avium*, all now reclassified in the new genus *Enterococcus* (Collins *et al.*, 1984; Schleifer & Kilpper-Bälz, 1984) and *S.equinus* and *S.bovis*. However, the recent revision of the genus *Enterococcus*, with 16 species recognized (Holt *et al.*, 1993), makes it necessary to revise the term faecal streptococci in order to establish which of these new species can be considered FS. Many laboratories give results of enterococci under the concept of faecal streptococci depending on the media used (Mundt, 1982; Pourcher *et al.*, 1991), and due to the fact that *S.bovis* and *S.equinus* survive shorter in aquatic environments than the enterococci (Geldreich & Kenner, 1969). Despite the opinion of certain authors that faecal streptococci are rarely found in an unpolluted environment (Rees, 1993) several studies seem to demonstrate that they can be found in other habitats or in plants and insects (Geldreich & Kenner, 1969; Mundt, 1982; APHA, 1992). This means that low counts for faecal streptococci even of around 100 cfu/100ml (only standard of EC Directive) cannot be exclusively interpreted as being of faecal origin (Geldreich, 1978). In addition, it has been also demonstrated that some of the strains of non faecal origin survive longer than those of faecal origin (Geldreich & Kenner, 1969). Several epidemiological studies have demonstrated adverse health effects at very low concentration from 20 to 35 FS/100ml (Cabelli *et al.*, 1982; Ferley *et al.*, 1989; Kay *et al.*, 1994).

In relation to the other microbial parameters, water samples are not normally examined routinely for *Salmonella* and enteroviruses. The absence of both *Salmonella* in 1 l of seawater and enterovirus in 10 l of water is in practice extremely difficult, almost impossible to achieve (Grantham, 1992; Rees, 1993). Grantham (1992) considered that an obvious anomaly in the Directive is that it requires zero enteric viruses yet it allows a level of enteric bacteria and therefore it would be impossible to guarantee compliance with this standard for viruses. Various studies have shown that especially enterovirus (and rotavirus) are ubiquitously present in bathing waters, often in the absence of bacterial indicators (Bosch *et al*, 1991; Rees, 1993). The same happens with *Salmonella*, it is present according to our experience in seawater 1 to 5% of the samples under guide limits for faecal coliforms. Other authors have also isolated

Salmonella under these conditions (Dutka, 1973; Slaider *et al.*, 1981; Moriñigo *et al.*, 1992). The previously commented point of the Directive presents the contradiction that those microorganisms are present when no reasons could make their presence suspect. The main contradictory aspect from our point of view is that both parameters are not mandatory for drinking water. Presence or absence is not a good approach for *Salmonella* either, a better one would be quantification and serotyping. What is the meaning of finding the presence of *Salmonella* with a serotype which is never present in clinical isolates? In fact, the detection of pathogens as parameters in seawater will only indicate their presence in the population, i.e. the degree of carriers or infected people in the population and the effectiveness of the sewage treatments and of the natural autodepuration factors such as dilution, stress etc. In this sense the inclusion of any pathogen in the Directive should be looked at carefully because their incidence changes from country to country. It seems that a general consensus exists in England rejecting the criteria for *Salmonella* and Enteroviruses (National Rivers Authority, 1990). Other authors consider that it would seem prudent to maintain some sort of viral standard for waters used for recreational purposes where total immersion is a feature of the activity (Morris, 1991).

2.2 Microbiological Methodology

Articles 9 and 10 reflect the possibility of modifications of the analytical methods and of their numerical standards for their adaptation to the technical progress. In this sense it is clear that the Directive requires an homogenization and updating of the methodologies.

The EC through the BCR (Bureau Communitarie de Référence) has sponsored and organized intercalibration exercises of microbiological methods (FC, FS) within the frame of the Measurement & Testing EC Programme for the analysis of seawater. Several representatives of the state members have participated. The final report will probably be very conclusive for changes to be introduced in the methodologies. This initiative of the Commission has to be evaluated very positively since it is no use to compare results of different countries if the procedures of detection and enumeration are not clearly standardized.

2.3 Homogenization Among Coutries

From our point of view it is also important that all countries evaluate the same parameters, both those that are mandatory and those that are recommended. Otherwise, when more controls are applied, not only the mandatory TC and FC, the possibility of failing is much higher mainly for *Salmonella* and Enteroviruses.

2.4 Comparison With Other Regulations

As commented by Kay *et al.* (1990) the EC guide level for TC and FC is the most stringent, while the mandatory standard is the least stringent, when compared with North American and Canadian standards. No other international regulation includes *Salmonella* and Enteroviruses.

3 EVOLUTION OF THE MICROBIOLOGICAL CONTROL IN THE EC

The EC Commission (DG XI) publishes a report each year on the quality of bathing waters based on the reports presented by all the Member States. These reports have attracted considerable attention from the Member States and general public. We will comment on the data that appeared in the most recent 1994 report (European Commission, 1995).

When evaluating the evolution of controlled sites in the EC in 1982, 7,000 sites were controlled, while 15,000 sites were controlled nine years later and more than 17,000 sites in 1994. The 12% increase of sampling points of the last four years is not homogeneously distributed among the countries. While some countries decreased the number of sites controlled (Germany, seawater -19%; Holland, continental - 46.7%) others increased them moderately e.g. Italy, Spain, France (10% to 22%) and others duplicated the controlled sites (Ireland) during this period.

Sampling mean frequencies ranged from 6 to 35 samples, the mean being 8.2 for continental waters and 12 for seawater. The lower frequencies agree with the fortnightly sampling established by the Directive over a 3 month summer bathing season while the middle frequencies indicate a week sampling régime over this period. An evaluation of the evolution of the microbiological control in relation with the obligatory TC and FC shows that most of the countries have controlled TC and FC for many years although Holland did not control TC until 1994 for continental water and until 1992 only 63% of the points for seawater.

The situation is slightly different for Faecal streptococci which received, in our opinion, considerable influence from the Blue Flag Campaign which has also demanded this parameter since 1992. In general, the countries that started in 1991 in a few points of control for FS increased the points progressively until 1994. In relation to *Salmonella* several countries such as Denmark, Greece, Luxembourg, have never evaluated this parameter, while others have been controlling this parameter since 1991 (Belgium, Germany, Spain, France, Ireland, Italy, Holland, Portugal, and the UK) and some have only carried out the test at a few points such as France and Holland (approx 3-6%). Others evaluated approximately 50% of points (Germany, continental; Spain) and others evaluated an even higher percentage of points (Belgium, Portugal, United Kingdom).

In relation with Enteroviruses most of the coutries do not control this parameter with the exception of Germany, Spain, France and the UK.

4 MONITORING PROGRAMME IN CATALUNYA

Some authors (Jones *et al.*, 1990) have commented that one of the major problems associated with the EC Directive is that they produce statistical information too late in the bathing season to enable any appropriated management action. They have also commented on the impossibility of providing meaningful statistical calculations with the small number of samples available during the bathing season. From our point of view to accomplish these objectives an intensive monitoring programme is required in close connection with the competent authorities that can take management decisions if necessary, and must guarantee regular information of the water quality to the population. In this sense the monitoring programme established in our region (Catalunya) fulfils all these aspects. The week-end sampling régime established, guarantees that every week users are informed through the media (TV, teletext, newspapers), about the water quality of each beach the week before (aesthetical aspects are also taken in account). Any change over the expected values of contamination at any beach is directly investigated, resampling and inspecting the site. In this sense, breaks of long sea outfalls have been detected and repaired in less than four days, as well as the presence of any uncontrolled sewage outfall. In those instances no apparent visual change of the colour or aspect of the water was detected.

It is clear that while bathing, people want water to be risk free of infections, but they also want it not to smell, look cloudy, taste peculiar and to be free of litter floating in it. The monitoring programme combines in the week valoration, the microbiological data and the aesthetical aspects of the sand and water obtained on a daily visit to the beach.

5 EVALUATION OF THE NEW PROPOSED BATHING WATER DIRECTIVE

The new proposal of Bathing Water Directive (EEC 94/C 112/03) has introduced several important changes while it has remained unchanged in other aspects.

5.1 Microbiological Parameters

The total coliforms have disappeared from the new proposal, this is in agreement with many suggestions as commented previously. The faecal coliforms have been replaced by *E. coli*, this is a good homogenization on the basis of previous comments. However, in practice the results of the Measurements & Testing EC Project demonstrate that no single method applicable to all types of waters within the EC that can produce homogeneous results for this parameter exists. In spite of the fact that the parameter has been changed the standards have remained which means that the Directive is more permissive since *E. coli* represents around the 77% of the FC on the basis of our results (Figueras *et al.*, 1994) and those of other authors (Freier & Hartman, 1987, Santiago-Mercado *et al.*, 1987).

Faecal streptococci have been included as an obligatory parameter in agreement with much research data as commented previously (Sinton *et al.*, 1993a,b). In relation with the standards, the guide level has been maintained reducing the percentage of compliance, since a mandatory level has been introduced (400 cfu/100 ml 95% compliance). However, it remains unclear from which scientific criteria this mandatory level has been chosen. Resampling for any unexpected increase of the parameter is established in the Directive. From our point of view resampling should be mandatory for all the parameters not only for this one, as is the case in other regulations (Kay *et al.*, 1990).

In relation with the pathogens previously included in the Directive *Salmonella* and Enteroviruses, the first one has disappeared but it is still named in Art.6-3, as a measure of diffuse contamination. Enteroviruses have remained and have become a parameter of obligatory evaluation, with the same previous standard, and a monthly frequency established.

The agents primarily involved in waterborne viral disease (hepatitis and gastroenteritis) are at present difficult to detect in water. Much information is available about entero and reoviruses but these viruses appear to be less important from an epidemiological point of view (IAWPRC,1991). In this review paper they suggest that the virological safety of water cannot at present be reliably assessed by direct analyses for viruses. If this parameter is going to stay in the final version of the Directive there is a need for a homogenization of the methodologies since otherwise results can be highly variable. The obligatory control for this parameter is in contradiction with reducing the financial pressure that the monitoring programmes produce on the member states.

The bacteriophages are included, although with the idea of replacing the enteroviruses by this parameter. No standard, nor a frequency and methodology are proposed. Numerous authors have suggested that colifages could be good indicators of faecal pollution (Borrego *et al.*, 1990; IAWPRC,1991), since they are less affected than bacteria by the physico-chemical factors in water purification (Abad *et al.*, 1994). These organisms may serve as both bacterial and virus indicators, as they show correlation with their bacteria host and similarities in behaviour to the pathogenic enteric viruses and their determination is cheaper. However, before establishing this parameter in the Directive, standardization of the phage-assay techniques on an international basis will be the first step to enhance the comparability of data from different laboratories.

5.2 Frequency And Validation Of Results

Many authors have already commented on the impossibility of providing meaningful statistical calculations with the small number of samples available during the bathing season and that sampling frequency deserves attention (Jones *et al.*, 1990; Philipp, 1991).

The new proposal has clarified the previous possibility of reducing the fortnightly sampling in a factor 2 when the microbiological quality of water in previous years was shown to be very good, defining what the required characteristics for the reduction are and considering which are excellent quality bathing areas with the objective of reducing financial cost. In our opinion it is an important fact that a clarification of criteria for reducing frequencies has been made. However, in our region, as in other Mediterranean countries, the high water temperature may allow bathers to obtain longer exposures and therefore the reduction of the frequency should only be recommended in colder climates (Northern European Countries). The reduction means one sample each month. This frequency will not allow the detection of problems fast enough to enable the timing of management decisions to be effective enough to accomplish the objectives of the Directive. Furthermore, the statistical significance of the results will be even lower. From our point of view the frequency reduction should be established not only depending on the excellent quality of the water but on the bathing population exposed.

The pass and fail rates now present in the Directive means 1 failure to fulfil the imperative standards over 20 discrete results. This failure can easily be due to rainfall during the preceding 24h (Grantham, 1992). Our experience, over 5 years of monitoring, in which 17 or 18 discrete results were obtained, indicates that in certain years we had the bad luck of sampling under these conditions, so the microbiological quality of water deteriorated.   The rainfalls were not heavy enough to be included in Art. 5.3, which enables the elimination of the data on the basis of climatological adverse situations. In our point of view resampling may be included when the quality of water has deteriorated which may be due to any degree of rainfall. If water quality has improved in the new sample, which is normally the case after 48h., this result is the one that should be considered. It is important that the population is aware that water quality can deteriorate considerably after heavy rain and that this general idea can prevent people from bathing on those days.

The results of the monitoring programmes are normally based on a single determination that can produce a high dispersion of results depending on the methodology employed (storage, multiple tube fermentation, membrane filtration, culture media employed, confirmation tests etc). The variability of the results has been previously mentioned, and it is considered that results should be treated as confidence intervals (Fleischer and McFadden, 1980). It is widely accepted that coliform bacteria concentrations in natural waters are best described as logarithmic normal probability distributions or by the geometric mean and not as simple percentages established in the Directive (Kay *et al.*, 1990), similar to the interpretation of the results in other international regulations (USEPA, 1986; Laws of Hong Kong, 1989; Kay *et al.*, 1990). Those approaches will better correct the individual deviations previously described.

5.3 Control Of Sources Of Contamination

Articles 6.3 and 6.4 are destined to stimulate the research of the causes that may produce the deterioration of water quality and reinforce the aspects pointed out at the earlier Directive. However, it is not obligatory to report those aspects to the commission. The inclusion of the excellent quality bathing areas comes into contradiction with this point, since any change over the expected values of contamination at any site will not be detected and investigated for 30 days and will produce information too late for any appropriate management action.

5.4 Definition Of The Bathing Area

This definition had remained as postulated in the previous Directive and has been claimed as imprecise and subject to widely differing interpretations (Howarth, 1992). In the words of the Directive (Art. 1.2a) *"bathing water is defined to mean all running or still fresh waters or parts thereof and sea water, in which: bathing is explicitly authorized by the competent authorities of the member state, or in which bathing is not prohibited and is traditionally practiced by a large number of bathers"*. In our view this concept is not homogeneous among member states: some coutries consider river outlets as sites of control while others exclude these sites. Surface waters are not considered by some state members as bathing water while used for recreational purposes. Many authors consider that the definition of bathing waters should consider the function of the water uses (bathing, sailing, diving) but apart from this we consider it important to consider the number of bathers exposed. Therefore, it is required that the Directive defines precisely what is a large number of bathers.

5.5 Sampling Specification

Details on sampling specifications such as the number of points on the function of the length of the beach, site (at the most polluted place, or at the least polluted one), depth (30 cm), state of the tide are not clearly established and must be pointed out to guarantee homogenization.

5.6 New Aspects

5.6.1 Public Information. Until now, apart from the EC yearly reports, the need for public information has not been included in the Directive. The new proposal includes the Art. 5.4, stating that the state members should ensure that information about the quality of the water is exposed at the site. This of course is a very important point that has already been considered by the Blue Flag Organization. Since it is not clearly established in which way the commission will control the application of this public information it can produce a non homogeneous application.

5.6.2 Prohibited Bathing Areas. The Directive established In Art.7, that when contamination may represent a danger to public health, the Member States will prohibit bathing at those sites. It is considered a danger when a significant deviation occurs in relation to the mandatory standards, and then the commission should be informed of the reason that produces such deteriorations. However, it is not clearly established what "a significant deviation" means. Therefore, the criteria can vary from country to country and in consequence this point should be numerically described.

6 INCIDENCE OF OTHER MICROORGANISMS IN BATHING WATERS

6.1 Other Coliform Genera

Many other microorganisms are present in bathing waters apart from *E. Coli*, among them other coliform genera: *Enterobacter*, *Klebsiella* and *Citrobacter* which include species considered as opportunistic pathogens to humans and therefore should not be underestimated (Figueras *et al.*, 1994).

6.2 *Salmonella*

In the frame of the EC Directive we have evaluated fortnightly the presence of *Salmonella* independent of considering that the quality of the water has deteriorated. The incidence in seawater samples was 9% in 1992 and decreased to 2.1% in 1993. The average incidence on beaches was 5.75%. In continental waters it ranged from 13.04% in 1993 to 8.3% in 1994.

We can conclude on the basis of our experience and from other data that the presence of *Salmonella* is highly influenced by rain due to the fact that storm water can come into contact with sewage, farm waste and soil runoff (Geldreich, 1978; Crane *et al.*, 1983).

Its presence (detection) is intermittent in a fortnightly sampling régime as well as when it is done each week; besides it is not always detected in places where its presence should be suspected. We agree with the findings of many authors that the percentage of presence is higher when the levels of coliforms increase. However, it was also found in 1-5% of samples with very low levels of coliforms (under guide levels or even in absence of indicators). This behavior is similar to that of many pathogens.

In our point of view, as commented previously, the presence/absence test does not help to clarify its possible sanitary risk much, if this parameter is not quantified and serotype. We also believe that probably its detection in absence of faecal contamination does not necessarily mean a high sanitary risk, at least for bathing.

6.3 *Aeromonas*

In recent years the interest on the pathological implication of *Aeromonas* (gastrointestinal disease, wound infections, systemic infection) has increased and reports of infections attributed to this microorganism following contact with water have appeared in recent years (Kuijper *et al.*, 1989; Kelly *et al.*, 1993). These aspects together with its high incidence in aquatic environments, and the unclear taxonomy of the group (Janda, 1991) have been the main reason that have led us to conduct a study to establish the frequency of the HGs (hybridization groups) in the environment in order to fully understand the epidemiology involved in *Aeromonas* infections.

The incidence of the main groups of *Aeromonas* and especially of the genospecies (following the phenotypic characterization proposed by Abbot *et al.*, 1992) has been evaluated. From our study we can conclude that the incidence of *Aeromonas* spp. in aquatic environments is very high, 96.8% continental waters and 88.6% seawater samples were positive. It has also been detected in 88.8% of continental waters and 53.8% of seawater samples in the absence of TC and FC. Continental waters have proved to be the main reservoir of those microorganisms due to the high diversity of genospecies encountered. The possible epidemiological significance of *A. veroni sobria* (often called *A. sobria*) has probably been underestimated (Borrell *et al.*, 1995). This species is 3rd in frequency in our study when a classical identification is applied.

However, it becomes the most frequent species, independently of the water source, on the basis of the phenotypic identification of genospecies. This species together with *A. hydrophila* has the highest virulence factors.

References

Abad FX, Pintó RM, Diez JM, Bosch A. Disinfection of human enteric viruses in water by copper and silver in combination with low levels of chlorine. Appl. Environ. Microbiol. 1994;60:2377-2383.

Abbott SL, Cheung WKW, Kroske-Bystrom S, Malekzadeh T, Janda JM. Identification of *Aeromonas* strains to the genospecies level in the clinical laboratory. J. Clin. Microbiol. 1992;30:1262-1266.

American Public Health Association. Standard methods for the examination of water and wastewater. 18th eds. Washington DC: American Public Health Association 1992.

Borrego JJ, Cornax R, Moriñigo MA, Martinez-Manzanares E, Romero P. Coliphages as an indicator of faecal pollution in water. Their survival and productive infectivity in natural aquatic environments. Water Res. 1990;24:111-116.

Borrell N, Figueras MJ, Guarro J. Phenotypical characterization of environmental *Aeromonas*. Proceedings of the 5th international *Aeromonas-Plesiomonas* symposium; 1995 April 8-9; Heriot-Watt University, Edinburgh.

Bosch A, Lucena F, Diez JM, Gajardo R, Blasi M, Jofre J. Waterborne viruses associated with hepatitis outbreak. J. Am. Water Works Assoc. 1991;83:80-83.

Cabelli VJ, Dufour AP, McCabe LJ, Levin MA. Swimming-associated gastroenteritis and water quality. Am. J. Epidemiol. 1982;115;606-616.

Cabelli VJ. Health effects criteria for marine recreational waters. U.S. Environmental Protection Agency. 1983; EPA 600/1, 80-031.

Collins MD, Jones D, Farrow JAE, Kilpper-Bälz R, Schleifer JH. *Enterococcus avium* nom. rev., comb. nov.; *Enterococcus casseliflavus* nom. rev., comb. nov.; *Enterococcus durans* nom rev., comb. nov.; *Enterococcus gallinarum* comb. nov.; and *Enterococcus malodoratus* sp. nov. Int. J. Syst. Bacteriol. 1984;34:220-223.

Council of European Communities concerning the quality of bathing water. 1975; (76/160/ECC). Official Journal L31:1-7.

Commission of European Communities. Proposal for a council directive concerning the quality of bathing waters. (94/C 112/03).

Crane SR, Moore JA, Grismer ME, Miner JR. Bacterial pollution from agricultural sources: A review. Trans. ASAE 1983;26: 858-866.

Dufour AP. *Escherichia coli*: The fecal coliform. In: Hoadley AW, Dutka BJ, editors. Bacterial indicators. Health hazards associated with water, 1977;635:48-58.

Dufour AP, Strickland ER, Cabelli VJ. Membrane filter method for enumerating *Escherichia coli*. Appl. Environ. Microbiol. 1981;41:1152-1158.

Dutka BJ. Coliforms are an inadequate index of water quality. J. Environ. Health 1973;36:39-40.

European Commission. Quality of bathing water 1994. (EUR 15976), Luxembourg, 1995.

Ferley JP, Zmirou D, Balducci F, Baleux B, Fera P, Larbaigt G. Epidemiological significance of microbiological pollution criteria for river recreational waters. Int. J. Epidemiol. 1989;18:198-205.

Fewtrell L, Jones F. Microbiological aspects and possible health risks of recreational water pollution. In: Kay D, editor. A perspective from the EC in Recreational water quality management. Ellis Horwood, Chichester, 1992:71-87.

Figueras MJ, Polo F, Inza I, Guarro J. Poor specificity of m-Endo and m-FC culture media for the enumeration of coliform bacteria in sea water. Lett. Appl. Microbiol. 1994;19:446-450.

Fleischer JM, McFadden RT. Obtaining precise estimates in coliform enumeration. Wat. Res. 1980;14:477-483.

Freier TA, Hartman PA. Improved membrane filtration media for enumeration of total coliforms and *Escherichia coli* from sewage and surface waters. Appl. Environ. Microbiol. 1987;53:1246-1250.

Geldreich EE. Applying bacteriological parameters to recreational water quality. J. Amer. Water Works Assoc. 1970;64:113-120.

Geldreich EE. Bacterial populations and indicator concepts in feces, sewage, stormwater and solid wastes. In: Berg G, editor. Indicators of viruses in water and food. Ann Arbor Science Publishers Inc., Ann Arbor, 1978:51-97.

Geldreich EE, Kenner BA. Concepts of fecal streptococci in stream pollution. J. Water Pollut. Con. F 1969;41(8 Part 2): 336-352.

Gilles V. The Bathing Water Directive. In: Kay D, editor. A perspective from the EC in recreational water quality management. Ellis Horwood, Chichester, 1992:19-24.

Grantham R. The role of the NRA in implementing the Bathing Water Directive. In Kay D, editor. Ellis Horwood, Chichester, 1992:25-31.

Holt JG, Krieg NR, Sneath PHA, Staley JT, Williams ST. Genus *Enterococcus*. In: Hensyl WR, editor. Bergey's Manual of determinative bacteriology. The Williams & Wilkins Co., Baltimore. 1993:528.

Howarth W. Legal issues concerning bathing waters. In: Kay D, editor. A perspective from the EC in Recreational water quality management. Ellis Horwood, Chichester, 1992:49-69.

IAWPRC Study Group on Health Related Water Microbiology: Bacteriophages as model viruses in water quality control. Water Res. 1991;25:529-545.

Janda JM. Recent advances in the study of the taxonomy, pathogenicity, and infectious syndromes associated with the genus *Aeromonas*. Clin. Microbiol. Rev. 1991;4:397-410.

Jones F, Kay D, Stanwell-Smith R, Wyer M. An Appraisal of the Potential Health Impacts of Sewage Disposal to UK Coastal Waters. J. IWEM 1990;4:295-303.

Kay D, Wyer M, McDonald A, Woods N. The application of water-quality standards to UK bathing waters. J. IWEM 1990;4:436-441.

Kay D, Wyer M. Recent epidemiological research leading to standards. In: Kay D, editor. A perspective from the EC in Recreational water quality management. Ellis Horwood, Chichester, 1992:129-156.

Kay D, Fleisher JM, Salmon RL, Jones F, Wyer MD and Godfree AF. Predicting likelihood of gastroenteritis sea bathing: results from randomized exposure. Lancet 1994;344:905-909.

Kelly KA, Koehler JM, Ashdown LR. Spectrum of extraintestinal disease due to *Aeromonas* species in Tropical Queensland, Australia. Clin. Infec. Dis. 1993;16:574-579.

Kuijper EJ, Bol P, Peeters MF, Steigerwalt AG, Zanen HC, Brenner DJ. Clinical and epidemiological aspects of members of *Aeromonas* DNA hybridization groups isolated from human feces. J. Clin. Microbiol. 1989;27:1531-1537.

Laws of Hong Kong. Water Pollution Control Ordinance. Statement of water quality objectives, Southern Junk Bay and Port Shelter water control zones. Government Printer, Hong Kong. 1989;358.

Moriñigo MA, Muñoz MA, Cornax, Martinez-Manzanares E, Borrego JJ. Presence of indicators and *Salmonella* in natural waters affected by outfall wastewater discharges. Wat. Sci. Tech. 1992;25: 1-8.

Morris R. The EC bathing water virological standard: Is it realistic? Wat. Sci. Tech. 1991;24: 49-52.

Mundt JO, Johnson O. The ecology of the streptococci. Microbial Ecol. 1982;8:355-369.

National Rivers Authority. Bathing water quality in England and Wales. Water quality series. 1990.

Philipp R. Risk assessment and microbiological hazards associated with recreational water sports. Rev. Med. Microbiol. 1991;2:208-214.

Pourcher AM, Devriese LA, Hernández JF, Delattre JM. Enumeration by a miniaturized method of *Escherichia coli, Streptococcus bovis* and enterococci as indicators of the origin of faecal pollution of waters. J. Appl. Bacteriol. 1991;70:525-530.

Rees G. Health implications of sewage in coastal waters - the British case. Mar. Pollut. Bull. 1993;26:14-19.

Saliba LJ, Helmer R. Health risks associated with pollution of coastal bathing waters. Wld. Health Stati. 1990;43:177-187.

Santiago-Mercado J, Hazen TC. Comparison of four membrane filter methods for fecal coliform enumeration in tropical waters. Appl. Environ. Microbiol. 1987;53:2922-2928.

Schleifer KH, Kilpper-Bälz R. Transfer of *Streptococcus faecalis* and *Streptococcus faecium* to the genus *Enterococcus* nom. rev. as *Enterococcus faecalis* comb. nov. and *Enterococcus faecium* comb. nov. Int. J. Syst. Bacteriol. 1984;34:31-34.

Sinton LW, Donnison AM, Hastie CM. Faecal streptococci as faecal pollution indicators: a review. Part I: Taxonomy and enumeration. N. Zeal. J. Mar. Freshwat. Res. 1993;27:101-115.

Sinton LW, Donnison AM, Hastie CM. Faecal streptococci as faecal pollution indicators: a review. Part II: Sanitary significance, survival, and use. N. Zeal. J. Mar. Freshwat. Res. 1993;27:117-137.

Slaider RJ, Evans TM, Kaufman JR, Warvick CE, Lechevalier MV. Limitations of standard coliform enumeration techniques. Research and Technology. J. Amer. Water Works Assoc. 1981:538-542.

UNEP/WHO. Determination of fecal coliforms in sea-water by the membrane filtration culture method. Reference Method. Mar. Pollut. Stud. 1983;2(Rev 1):1-19. UNEP. Geneva.

United States Ambient Water Quality Criteria for Bacteria. 1986. EPA 440/5-84-002. 18 pp.

United States Environmental Protection Agency. Health effects criteria for marine recreational waters. 1986. USEPA, Washington, D.C.

Chapter 13
Coliforms And *E. Coli* In Finnish Surface Waters

R. Maarit Niemi[1], Seppo I. Niemelä[2], Kirsti Lahti[1] and Jorma S. Niemi[1]

[1]   Finnish Environment Agency P. O. Box 140 FIN-00251 Helsinki Finland
[2]   Department of Applied Chemistry and Microbiology, University of Helsinki P.O. Box 27 FIN-00140 Helsinki University Finland

1 INTRODUCTION

Coliform bacteria and *E. coli* are discharged into surface waters from point and diffuse sources. The main point sources are treated domestic and industrial wastewaters, whereas diffuse sources include soil, vegetation and untreated wastes of humans in sparsely populated areas without sewerage systems, agricultural areas, cattle grazing and wild animals. The concentrations and species composition of the coliform flora observed in waters depend on the source of bacteria, sewage treatment process, dilution conditions and survival and regrowth conditions.

Coliform flora consist of a variety of species. *Escherichia coli* is regarded as a true faecal bacterium even if it has been reported to multiply when pure cultures have been exposed to the aquatic environment (Hazen 1988). In addition to *E. coli*, *Citrobacter freundii*, *C. diversus*, *C. amalonaticus*, *Klebsiella pneumoniae*, *K. oxytoca*, *K. mobilis* and *Enterobacter cloacae* are regarded as faecal bacteria (Gavini *et al.*, 1985). On the other hand, *Serratia fonticola*, *Rahnella aquatilis*, *Enteroba-cter intermedium*, *E. amnigenus*, *E. gergoviae*, *E. sakazakii*, *Buttiauxella agrestris*, *Klebsiella trevisanii*, and *K. terrigena* are described as environmental saprophytic species (Gavini *et al.*, 1985). Even if *Hafnia alvei* does not produce acid from lactose it is often isolated as a coliform bacterium. It is distributed in humans (occasionally as a pathogen), in animals and birds and in natural environments such as soil, sewage, and water (Sakazaki and Tamura, 1991).

Finland is situated between the 60th and 70th latitudes north. Most of the population and industrial activity are in the south whereas the northern part is sparsely populated and less industrialized. Seasonal climatological variations are large and there are all types of surface waters: slightly polluted waters in pristine areas and in non-point loading areas and waters which are heavily polluted by domestic and industrial wastewaters. On the basis of input of faecal bacteria and climate the country forms a south-north gradient. Because of these characteristics Finland is a suitable area for investigating the input and fate of coliform bacteria and *E. coli*. The overall hygienic quality of Finnish inland waters was investigated by Poikolainen *et al.* (1995). They found that since the 1960s the concentrations of thermotolerant coliform bacteria had decreased simultaneously with the construction of wastewater treatment plants.

This paper summarizes the occurrence of coliforms and *E. coli* in Finnish domestic wastewaters, industrial effluents and in surface waters of pristine, diffuse and point loading areas and evaluates their suitability as indicators of faecal pollution.

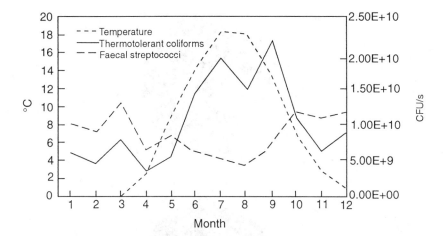

Figure 13.1 *Loads of total coliform bacteria and faecal streptococci discharged through the domestic wastewaters from the town of Kouvola to the River Kymijoki. (Difference of monthly geometric means between the bacterial loads upstream and downstream of the treatment plant).*

especially that of *Klebsiella* (Figure 13.2, $X^2_5 = 79.8$, P <0.001) increased. When holiday periods were excluded from the data the differences between the thermotolerant coliform florae of the influent and effluent wastewater were not significant ($X^2_3 = 3.19$; 0.05 <P< 0.10). This implies that the thermotolerant coliform flora had also changed before entering the treatment plant. The change was probably due to different die-off rates or to simultaneous multiplication of some species, indicating that in general other coliforms than *E. coli* found the sewage system a more favourable place for multiplication. Some indicator bacteria were found to multiply in a kitchen drain trap, e.g. *E. coli* showed a ninefold multiplication (Figures 13.3 and 13.4). In sewage, however, the concentrations of *E. coli* decreased in comparison to other coliform species.

At the activated sludge treatment plants the average annual reduction of total and thermotolerant coliform bacteria was about 95%.

3.2 Industrial effluents

Industrial effluents, which are rich in organic matter are known to interfere in the monitoring of faecal contamination (Vlassoff, 1977). There are characteristic differences in the composition of total coliform flora isolated from different point sources (Figure 13.5). In domestic wastewaters *E. coli* dominated followed by *Klebsiella*, whereas in forest industry wastewaters *Klebsiella* and *Enterobacter* were the most frequently isolated species. In textile and food industry wastewaters *Klebsiella*, *Enterobacter* and *Citrobacter* were commonly isolated. In fish farm effluents *Aeromonas hydrophila* and *Citrobacter* were the most common species. In industrial wastewaters a relatively high proportion of isolates remained unidentified. At fish farms total coliforms consisted of different genera, whereas of the 287 thermotolerant coliform isolates 89% were *E. coli*. This reflects faecal contamination due to e.g. bird droppings (Niemi, 1985).

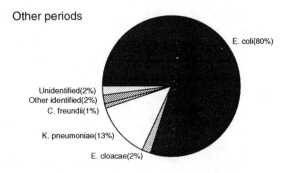

Other periods

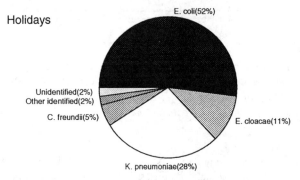

Holidays

Figure 13.2 *Composition of thermotolerant coliform flora (mFC T) at the wastewater treatment plant during holidays (n=193) and other periods (n=481).*

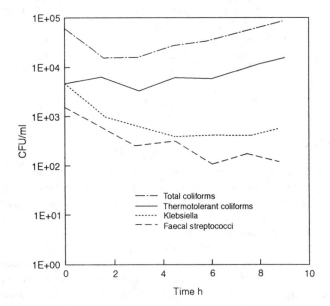

Figure 13.3 *Growth of faecal indicator bacteria in a kitchen drain trap at 22 °C. (Klebsiella were enumerated on mFC agar).*

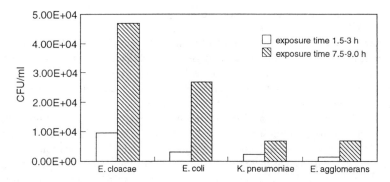

Figure 13.4 *Growth of different coliform species in a kitchen drain trap (n=109).*

Some industrial effluents have high concentrations of thermotolerant coliform bacteria and they interfere seriously in the monitoring of faecal contamination of waters, especially if the discharged volumes are high. In Finland an example of this is the effluents of forest industry. Niemelä and Väätänen (1982) studied the survival of *Klebsiella* discharged from forest industry to the receiving water body. At 0°C the decimal reduction time was about 24 days and at 20°C slightly over 5 days. *E. coli* seems to be suitable for the monitoring of faecal contamination of waters loaded with industrial effluents with one exception: it was observed to multiply persistently and in remarkable amounts in one pulp and cardboard mill with a unique process (Niemi *et al.*, 1987). Isolates from this mill differed serologically from human pathogenic and faecal strains. Even *E. coli* is not a suitable indicator of faecal contamination in the waters affected by these effluents.

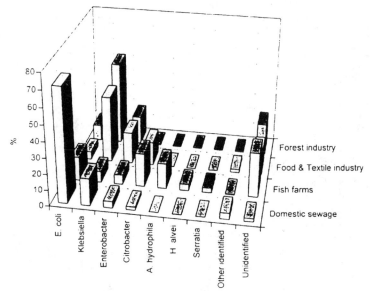

Figure 13.5 *Composition of coliform flora in different wastewaters. (Number of isolates: domestic sewage 674, food and textile industry 331, forest industry 359 and fish farms 1244)*

3.3 Waters in pristine, diffuse and point loading areas

The occurrence and species composition of total and thermotolerant coliform flora was studied in small brooks, rivers and small lakes situated in catchment areas with no human settlements, agriculture, industry or wastewater technology (Niemelä and Niemi, 1989). The northern study area lies at the limit of coniferous trees in the reindeer grazing region (about 69°N, 27°E). The southern study area is situated at an upland lake district (about 60°N, 24°E). A total of 372 coliform strains isolated from these two areas were identified (Figure 13.6). Non-faecal species clearly outnumbered *E. coli*. The observed total coliform counts, which varied from 80 to 11000 CFU/100 ml, misindicated therefore faecal contamination by humans or homoiothermic animals. All of the 44 thermotolerant coliform isolates from the southern region and 41 of the 42 thermotolerant coliform isolates from the northern region were presumptive *E. coli*. Thermotolerant coliform bacteria indicated reliably the occurrence of *E. coli* both in pristine and agricultural areas. In pristine areas 96% of 215 isolates and in agricultural areas 91% of 391 isolates were *E. coli* (Niemi and Niemi, 1991). Almost total absence of *Klebsiella* at the fish farms and pristine waters was the other remarkable feature.

In rivers the concentrations of thermotolerant coliforms are affected mainly by seasons (Figure 13.7), although rains can cause short term variations (Niemi and Niemi, 1988). In winter the riverine concentrations of thermotolerant coliforms and *E. coli* were high due to cold water, low light intensity and low dilution under ice cover. The concentrations were high also in spring when increased runoff due to melting snow washed faecal material into rivers and cold water promoted survival. In summer the bacterial concentrations were low, because high water temperature increased bacterial die-off. After heavy rains however the concentrations rapidly rose due to increased run-off (Niemi and Niemi, 1990).

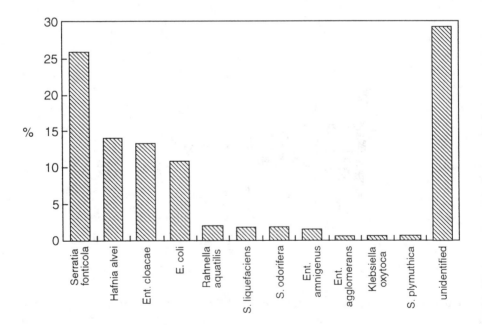

Figure 13.6 *Species composition of total coliform flora in pristine waters (n=372).*

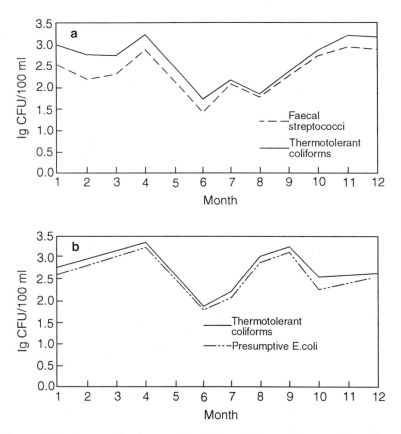

Figure 13.7 *An example of seasonal variation of thermotolerant coliform bacteria and presumptive E.coli in two rivers. a) The River Aurajoki: monthly averages of 8 successive years in one observation site. b) The River Vantaanjoki: monthly averages of 9 observation sites during one year. Concentrations expressed in lg CFU/100 ml*

Niemi and Niemi (1991) compared the concentrations of thermotolerant coliforms and *E. coli* enumerated from waters of small drainage basins in pristine and agricultural areas and from treated domestic sewage in summer (Table 13.2.). The results showed that point loading had the major impact on the hygienic quality of waters, although waters in agricultural areas could also be contaminated. Pristine waters tended to be of good hygienic quality, but were occasionally contaminated by wild animals.

4 CONCLUSIONS: COLIFORMS AS INDICATORS OF FAECAL CONTAMINATION

Total coliform bacteria are not suitable indicators of faecal contamination in surface waters, because of the frequent occurrence of environmental coliform organisms interfering in the

Table 13.2 *Statistical parameters of the concentrations of thermotolerant coliform bacteria and E. coli in pristine and agricultural areas and in treated wastewater (CFU/100 ml; Niemi and Niemi, 1991).*

Indicator	Parameter	Pristine area	Agricultural area	Treated wastewater
Thermotolerant	n	49	24	24
coliforms	median	0	307	17800
	min	0	8	500
	max	268	38200	414300
Presumptive	n	47	24	23
E. coli	median	0	283	9100
	min	0	7	300
	max	259	38200	100000

measurement even in pristine areas. Thermotolerant coliforms measured mainly faecal *E.coli* in pristine environments, in non-point loading areas without sewage systems and in fish farm waters, but in domestic wastewaters they were not equally reliable indicators because of *Klebsiella*. In industrial wastewaters high concentrations of thermotolerant nonfaecal coliform species interfered seriously with the detection of *E. coli*. In surface waters the enumeration of total coliforms and thermotolerant coliforms should be replaced by the enumeration of *E. coli* as methods for presumptive *E. coli* are now available. This would save labour and costs and indicate faecal contamination more reliably. Marked and prolonged multiplication of *E. coli* outside humans and homoiothermic animals is rare at least in our northern environments and these special cases can be studied separately.

References

1 Dufour AP. *Escherichia coli*: The fecal coliform. In: Hoadley AW, Dutka BJ, editors. Bacterial Indicators/Health Hazards Associated with Water.1977: 48-58. American Society of Testing and Materials, STP 635, Philadelphia.

2 Gavini F, Leclerc H, Mossel DA. Enterobacteriaceae of the "Coliform Group" in drinking water: identification and worldwide distribution. System.Appl.Microbiol.1985; 6:312-318.

3 Hazen TC. Faecal coliforms as indicators in tropical waters: a review. Toxic. Assess. 1988;3:461-477.

4 Leclerc H, Mossel DA, Trinel PA, Gavini F. Microbiological monitoring - a new test for fecal contamination. In: Hoadley AW, Dutka BJ, editors. Bacterial Indicators/Health hazards Associated with water. 1977:23-36 American Society of Testing and Materials, STP 635, Philadelphia.

5 Niemelä SI, Niemi RM. Species distribution and temperature relations of coliform populations from uninhabited watershed areas. Toxic. Assess. 1989;4:271-280.

6 Niemelä SI, Väätänen P. Survival in lake water of *Klebsiella pneumoniae* discharged by a paper mill. Appl. Environ. Microbiol. 1982;44:264-269.

7 Niemi RM. Fecal indicator bacteria at freshwater rainbow trout (*Salmo gairdneri*) farms.

Publications of the Water Research Institute 1985;No.64. National Board of Waters, Helsinki.

8 Niemi RM, Niemelä SI, Mentu J, Siitonen A. Growth of *Escherichia coli* in a pulp and cardboard mill. Can. J. Microbiol. 1987;33:541-545.

9 Niemi RM, Niemi JS. Annual variation and reliability of fecal indicators in a polluted river. Toxic. Assess. 1988;3:657-677.

10 Niemi RM, Niemi JS. Monitoring of faecal indicator bacteria in rivers on the basis of random sampling and percentiles. Water,Air, Soil Pollut. 1990;50: 331-342.

11 Niemi RM, Niemi JS 1991. Bacterial pollution of waters in pristine and agricultural lands. J. Environ. Qual. 1991;20:620-627.

12 Poikolainen ML, Niemi JS, Niemi RM, Malin V. Fecal contamination of Finnish fresh waters in 1962-1984. Water, Air, Soil Pollut. 1995;81:37-47.

13 Sakazaki R, Tamura K. The genus Hafnia. In: Balows, A, Tr.per HG, Dworkin M, Harder W, KH Schleifer (eds.) The prokaryotes. A handbook on the biology of bacteria: Ecophysiology, isolation, identification, applications 1991. 2nd ed., vol. III:2816-2821. New York.

14 Vlassoff LT. *Klebsiella*. In: Hoadley AW, Dutka BJ, editors. Bacterial Indicators/Health Hazards Associated with Water. 1977:275-288. American Society of Testing and Materials, STP 635, Philadelphia.

Chapter 14
Non-Sewage Derived Sources Of Faecal Indicator Organisms In Coastal Waters: Case Studies

M. D. Wyer[1] , J. G. O'Neill[2], V. Goodwin[2], D. Kay[1], G. Jackson[3],
L. Tanguy[3] and J. Briggs[2]

[1] Centre for Research into Environment and Health, The Environment Centre, University of Leeds, Leeds, West Yorkshire, LS2 9JT
[2] Yorkshire Water Ltd, Leeds, LS1 4BG
[3] States of Jersey Public Services Department, Jersey, CI

1 INTRODUCTION

The application of modern sewage disposal methods, including long sea outfalls and tertiary treatment of final effluent, has improved the quality of coastal bathing waters as measured by concentrations of faecal indicator organisms. Examples include long sea outfalls constructed to protect bathing beaches on the Yorkshire coast in the past five years. The dilution and dispersion of effluent, and subsequent bacterial die off, from these outfalls has significantly improved water quality at identified locations monitored by the National Rivers Authority (NRA) and assessed in relation to the bacteriological parameters of the EC Directive 76/160/ EEC for bathing waters, summarised in Table 14.1 (EC, 1976).

A particularly good example is Flamborough on the Yorkshire coast. Prior to the construction of the long sea outfall at Flamborough South Landing, primary settled sewage table from the town of Flamborough discharged to a dyke which cascaded over a cliff at a point 400m from the designated bathing beach. Figure 14.1 shows how this small beach regularly failed the *Imperative* standards of the EC Directive until construction of the long sea outfall in 1992. After construction of the outfall the beach has achieved EC *Guideline* levels for total and faecal coliforms and faecal streptococci.

The States of Jersey Public Services Department (PSD) have adopted an alternative disposal scheme, involving the disinfection of secondary treated effluent from the island sewage treatment works using ultraviolet radiation. This is the first full scale UV treatment plant in Europe and discharges to St Aubin's Bay at the mean half tide level via a short outfall (Figure 14.2). The high quality effluent from the plant has dramatically reduced the indicator organism load from the treatment works since 1991.

Table 14.1 *EC Directive Standards for indicator organisms (EC, 1976)*

| Organism | Standard | |
	Imperative	*Guide*
Total coliforms	95% < 10,000	80% < 500
Faecal coliforms	95% < 2,000	80% < 100
Faecal streptococci	--	90% < 100

units = cfu 100ml^{-1}

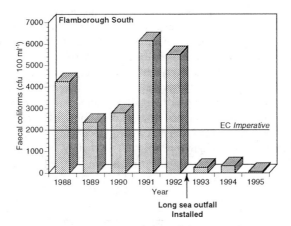

Figure 14.1 *Results of compliance monitoring (expressed as the second highest faecal coliform concentration (cfu 100 ml⁻¹) from 20 samples) at Flamborough South Landing beach on the Yorkshire coast, 1988-1995*

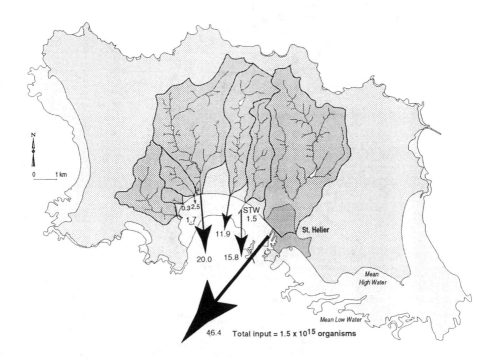

Figure 14.2 *Estimated % contributions in runoff to the faecal coliform input to St. Aubin's Bay, Jersey*

In comparison to bacterial loads from streams draining to St Aubin's Bay, the final effluent from the treatment works has been estimated to account for just 1.5% of the faecal coliform load (Figure 14.2) whilst the effluent stream provides almost 50% of the freshwater input to the bay. Total coliform loads showed a similar pattern (Wyer *et al.*, 1995b).

Despite the completion of such capital works to improve water quality relatively high concentrations of faecal indicator organisms can been found in bathing waters. In some instances indicator concentrations have been high enough to produce failure to comply with the EC Bathing Water Directive. An illustration of this comes from Scarborough, where water quality at the North beach showed a significant improvement after the installation of a long sea outfall, similar to Flamborough (Figure 14.3a). In contrast, Scarborough South beach has not shown such a pronounced improvement in relation to the EC standards until recently (Figure 14.3b).

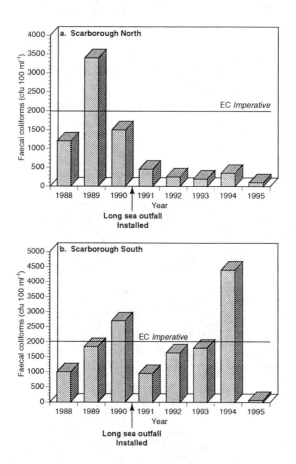

Figure 14.3 *Results of compliance monitoring (expressed as the second highest faecal coliform concentration (cfu 100 ml⁻¹) from 20 samples) at Scarborough North and South beaches on the Yorkshire coast, 1988-1995.*

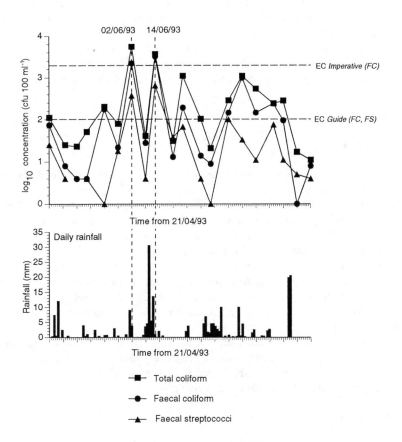

Figure 14.4 *Indicator organism concentrations (cfu 100 ml⁻¹) at First Tower St Aubin's Bay, Jersey, compared with daily rainfall, 1993*

In Jersey, bathing water quality in St Aubin's Bay has shown compliance failures which relate to rainfall events and corresponding elevations in bacterial concentrations in streams (Figure 14.4) (Wyer *et al.*, 1994).

The lack of dramatic improvement at a beach following capital investment in an outfall or treatment system is not uncommon. It is obviously very important to evaluate the actual effect of the outfall / treatment system on a beach, as distinct from modelling predictions. Results from Yorkshire and Jersey suggest that other sources of faecal indicators may be under estimated in relation to compliance with bathing water quality standards. In some cases, these other sources of contamination may only be revealed after the reduction of the main input (in the case of UV treatment), or its effect (in the case of a long sea outfall) has taken place.

Obvious possible sources of contamination include storm water overflows (which have been known to operate in the absence of storms) and river and stream discharges. Sewage cannot always be ruled out, either indirectly via a water course or as a direct source of beach contamination. The integrity and even the location of sewerage is not always readily defined and systems adjacent to water courses or beaches must always be considered as potential sources of contamination even in the absence of any apparent surface flow. However, this paper concentrates on a series of examples involving non-sewage related sources of faecal indicators which have impacted on or have a potential to affect water quality at bathing beaches.

2 CASE STUDIES

2.1 Case Study A

This case study was part of an investigation into possible sources of contamination of a beach. An effluent, initially thought to be a combination of sewage and an innocuous trade effluent, was found to discharge at the low water mark, not continuously but for 5-10 minutes four times an hour. The absence of any sewage content in the discharge was eventually demonstrated and the discharges formed visible plumes which could clearly be seen moving parallel with the beach following the longshore current.

In the food processing plant low numbers of faecal indicators could be detected in the raw material (Table 14.2), probably from small mammals. These appeared to grow during a steeping process, although the temperature did not exceed 20°C. The discharge was essentially the water used to wash the material after steeping for 24 hours. This wash water was held in a tank where further growth could have taken place.

Although up to 2500 m^3 per day could be discharged from the plant, it was concluded from tracing and intensive sampling that the impact on the beach was well below that required to exceed the mandatory standards of the Directive. The important factor was almost certainly the initial dilution of the effluent on contact with the sea.

2.2 Case Study B

The second case study presented involved another food processing effluent discharging directly to sea, complicated by the fact that the pipe could also act as a storm water overflow for a sewerage system. However, a characteristic of this discharge was that numbers of faecal indicators, although highly variable, could be greater than those in raw sewage. Also, faecal coliforms to faecal streptococci ratio (FC:FS) in the effluent was significantly different to that of raw sewage. As in case study A the high counts of faecal indicators arose from growth within the factory. In this case, water close to boiling point was used for washing the raw material. This water was constrained in an open channel within the factory with sufficient

Table 14.2 *Typical values of faecal indicator concentrations (cfu 100ml^{-1}) from food processing plant A*

	Source	
Indicator	Raw Material	Washwater discharge
Total coliforms	2.3x10^2	6.0x10^5
Faecal coliforms	2.0x10^1	5.0x10^3

Table 14.3 *Faecal indicator concentrations (cfu 100 ml⁻¹) and FC:FS ratio from food processing plant B*

Indicator	Median	Maximum
Total coliforms	2.3×10^8	9.6×10^8
Faecal coliforms	9.5×10^7	1.4×10^8
Faecal streptococci	4.0×10^8	1.4×10^9
FC:FS ratio	0.238	0.100

temperature and time (although not easily defined) for extensive growth of faecal indicators, which are probably soil derived.

When compared to raw sewage, the concentrations of the two coliform groups were around one order of magnitude greater in this discharge (Table 14.3). Concentrations of faecal streptococci were two orders of magnitude greater than in raw sewage.

In a particular year a designated beach 2 km away from the plant failed to comply with the EC Directive standards. This beach generally had low faecal indicator organism counts and a FC:FS ratio above unity (Table 14.4). However, a detailed analysis of the FC:FS ratio in those samples which produce compliance failure (i.e. the two samples with the highest indicator organism concentrations) showed these samples to have a relative elevation in faecal streptococci concentration similar to the plant effluent (Table 14.4).

2.3 Case Study C

Although initially unexpected as sources of faecal indicators, the two previous sources were relatively easily defined. Case study C represents a much less identifiable source of faecal indicators. Again, the growth of faecal indicators was eventually identified. The beach under investigation is no more than 30m from a harbour where fish are landed. The activity involved was the washing down of boxes in which the fish were packed. The wash water itself contained no detectable coliforms but this ran away through a drainage system into the sea on the beach side of the harbour wall. This washing resulted in biofilm slime accumulation in the internal drainage system, providing a substrate for growth. When washing took place the wash water drew out sufficient numbers of bacteria to affect the bathing beach (Table 14.5).

The concentrations involved were not high enough to be solely responsible for compliance failure at the beach. However, this source in combination with other sources of contamination

Table 14.4 *The two highest faecal indicator concentrations (Sample 1, Sample 2), median values (cfu 100ml⁻¹) and FC:FS ratio in NRA samples taken at the beach receiving effluent from food processing plant B*

Indicator	Sample 1	Sample 2	Median
Total coliforms	1.6×10^4	3.5×10^4	1.1×10^2
Faecal coliforms	1.1×10^4	8.3×10^3	3.0×10^1
Faecal streptococci	9.7×10^4	8.8×10^5	2.0×10^1
FC:FS ratio	0.111	0.009	1.500

Table 14.5 *Typical values of faecal indicator concentrations (cfu 100 ml⁻¹) in drainage to the beach from washing fish packing boxes (case study C)*

Indicator	Concentration
Total coliforms	1.6×10^5
Faecal coliforms	1.6×10^5
Faecal streptococci	2.2×10^5

typical of beaches close to urban areas can be considered an important contributory factor to failures at the beach.

2.4 Case Study D

Case study D involves a popular bathing beach remote from a sewage effluent source. Despite this remoteness, marine water quality often exceeds the EC *Guide* standard criteria and can approach the *Imperative* standard level (Figure 5). Observations suggest that marine water quality deteriorates during periods of high rainfall when discharge from a natural stream draining to the bathing water is high. During high flow response to rainfall the faecal indicator organism concentrations at the stream outlet to the beach were significantly elevated (Figure 6), a pattern observed in many other streams discharging to the coastal zone (Wyer *et al.*, 1994). Elevations in faecal indicator organism concentrations at the outlet compared to baseflow

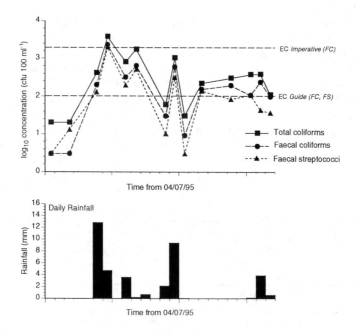

Figure 14.5 *Faecal indicator concentrations in marine samples from beach D and daily rainfall, July 1995*

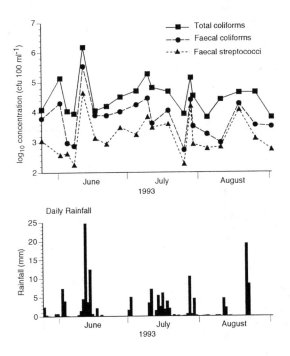

Figure 14.6 *Faecal indicator concentrations in the stream outlet to beach D and daily rainfall, summer 1993*

were: TC 54x to 1.5×10^6 100 per ml, FC 67x to 3.5×10^5 per 100 ml and FS 35x to 4.5×10^4 per 100 ml.

Figure 7 shows the results of a survey of water quality in the catchment contributing to the discharge at the beach. Samples were taken at five sites on the main stream, site 1 being the stream outlet and site 5 being 2.25 km upstream, in the catchment headwaters. The survey revealed high indicator organism concentrations downstream of captive water fowl populations resident in the catchment near sites 2 and 4 (Figure 14.7). Each individual bird can produce more *E. coli* per day (1.11×10^{10}) than a human (Jones and White, 1984) and their enclosures encompassed the natural stream channel. Organisms excreted by the water fowl are thus likely to be flushed to the coastal water at the beach during hydrograph response to rainfall events, producing a corresponding deterioration in bathing water quality (Wyer *et al.*, 1995a). In this example management of faecal indicator organism sources in the catchment is likely to produce a sustainable improvement in water quality at the beach.

2.5 Case study E

This study concerns another small beach unaffected by sewage effluent. Water quality monitoring at the beach revealed excellent water quality in 1991, most indicator organism

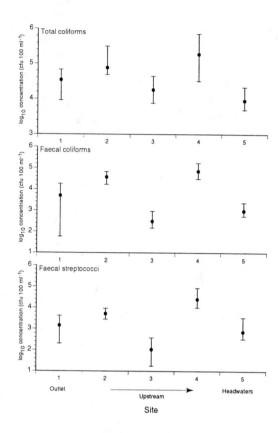

Figure 14.7 *Geometric mean and range faecal indicator concentrations (cfu 100 ml⁻¹) in the stream draining to beach D, summer 1993. (Sites 2 and 4 are downstream of captive water fowl)*

counts being below a detection limit of 10 cfu 100 ml⁻¹, well within the EC *Guideline* Standards (Figure 14.8).

Compliance monitoring in 1992 revealed a significant deterioration in water quality (Figure 14.9) compared to the previous bathing season. Indicator organism concentrations in the sea were within the EC *Guideline* criteria for the first half of the bathing season until July. Several samples subsequently exceeded the *Guideline* criteria whilst one exceeded the *Imperative* criteria. The deterioration in water quality at this normally exceptionally clean beach was directly related to the burial of 4000 tonnes of surplus potatoes in a field 300m from the beach during early July 1992.

The decomposing potatoes (which contain 95% water) produced a malodorous leachate draining directly to the beach via a natural spring and stream channel at up to 70 m³ per day. This leachate had a chemical oxygen demand (COD) of 24-36000 mgl⁻¹ and a pH between five and six. Water within the sand at the beach was turned anaerobic with a sulphide

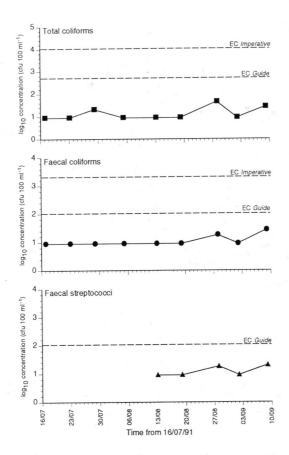

Figure 14.8 *Faecal indicator organism concentrations (cfu 100 ml⁻¹) at beach E, 1991*

concentration of 690 mgl⁻¹ H_2S. This decomposing organic matter provided an ideal substrate for the growth of faecal indicator organisms, whilst the pH was relatively high, which were discharged to the sea in the leachate at concentrations enough to produce an exceedance of EC *Guideline* and *Imperative* criteria. Subsequent acidification of the leachate ensured that the associated microbial bloom was transient.

Remedial capital works have since been completed to catch and store the high COD leachate for disposal into the foreseeable future. Water quality at the beach subsequently returned to within the EC *Guideline* limits (Figure 14.10)

2.6 Case Study F

The final study concerns a redundant quarry used for the disposal of high water content "green waste" that cannot be incinerated. The waste deposited at the site includes seaweed

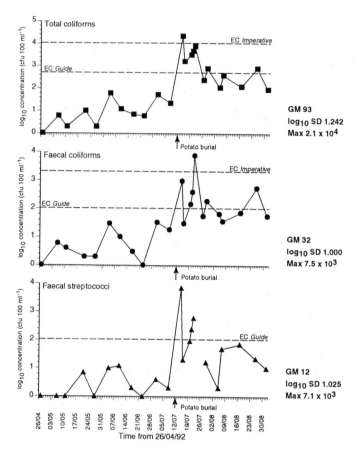

Figure 14.9 *Faecal indicator organism concentrations (cfu 100 ml⁻¹) at beach E, 1992*

regularly removed from beaches during the bathing season and the contents of road cleaning wagons. The latter will provide an inoculation of faecal material to the site, from sources such as dog faeces.

The leachate from the site is foul smelling, very turbid and brown coloured. The liquor drains from the site via a concrete drainage channel, finally discharging to sea via a pipe. Geometric mean concentrations of all three faecal indicator organisms in the leachate are high (TC 3.0×10^6 cfu 100ml⁻¹, FC 3.0×10^5 cfu 100ml⁻¹, FS 5.8×10^4 cfu 100ml⁻¹), maximum concentrations approaching values found in raw sewage (Figure 14.11).

Whilst this discharge does not appear to affect the nearest monitored beach, which enjoys excellent water quality, such a site could potentially affect marine water quality in a different location.

3 CONCLUSIONS

The range of non-sewage related sources of faecal indicators explored in this paper demonstrate the importance of understanding alternative sources of faecal indicator organisms

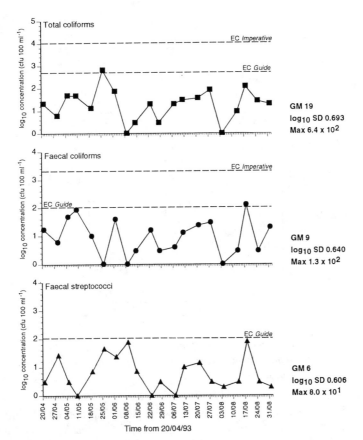

Figure 14.10 *Faecal indicator organisms concentrations (cfu 100 ml⁻¹) at beach E, 1993*

when assessing the potential for a site to comply with standard systems based on such indicators. Assessments of the likelihood of compliance with Directive standards will increasingly require research into, and the evaluation of, such sources if sustainable improvements in water quality at bathing beaches are to be achieved. As the case studies illustrate, such sources are often obscure and only indirectly related to faecal matter. In some cases remedial actions can be simple and expensive.

An understanding of such sources is likely to be important to avoid unnecessary capital expenditure at locations where more obvious discharges, such as sewage effluent, may not be the single cause of compliance failure. However, research experience in this field has demonstrated that such sources of faecal indicators are often only revealed after capital expenditure to improve obvious discharges has taken place.

Acknowledgements

The opinions expressed in this paper are those of the authors and not necessarily those of the organizations represented.

We thank Duncan Berry, Lee Butcher, Claire Le Breuilly and Kate Little of the States of

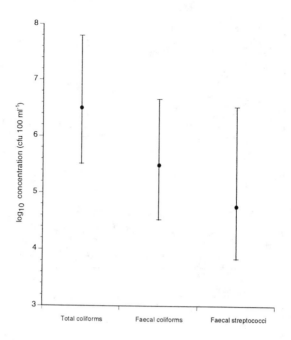

Figure 14.11 *Geometric mean and range of indicator organism concentrations (cfu 100 ml⁻¹) in leachate from case study F, a 'green waste' (e.g. seaweed, vegetable matter) disposal site*

Jersey Public Services Department and Helen Dawson, Adrian Spence, John Whittle and Jan Yeo of Acer who undertook many of the sample analyses. We also thank Angela Duignan and Ben Woods of the Environment Centre, University of Leeds, who assisted with sample collection and analysis.

References

EEC (European Economic Community) 1976. "Council of 8 December 1975 concerning the quality of bathing water (76/160/EEC)." *Official Journal of the European Communities* **L31**: 1-7.

Jones F. and White W.R. 1984. "Health and Amenity aspects of surface waters." *Water Pollution Control* **83**: 215-225.

Wyer M.D., Jackson G.F., Kay D., Yeo J. and Dawson H. (1994). An assessment of the impact of inland surface water input to the bacteriological quality of coastal waters. *Journal of the Institution of Water and Environmental Management, 6,* 459-467.

Wyer M.D., Kay D., Jackson G.F., Dawson H.M., Yeo J. and Tanguy L. (1995a). Indicator organism sources and coastal water quality: a catchment study on the Island of Jersey. *Journal of Applied Bacteriology , 78,* 290-296.

Wyer M.D., Crowther J. and Kay D. (1995b). *Further assessment of non-outfall sources of bacterial indicator organisms to the coastal zone of the Island of Jersey.* A report to the Public Services Department of the States of Jersey. 27pp. CREH, The Environment Centre, University of Leeds.

Chapter 15
Strategies For Survival

Keith Jones

Division of Biological Sciences, I.E.B.S., Lancaster University, Lancaster LA1 4YQ

1 INTRODUCTION

Escherichia coli has a 'boom or bust' existence which depends on its ability to grow rapidly, when nutritional and environmental conditions are good, and to remain viable during nutrient starvation and adverse conditions. *E. coli* and the other coliform bacteria have evolved a sophisticated programme of physiological and morphological changes which they undergo when they enter the stationary phase of the growth cycle. The modified cells are resistant to a wide range of environmental stresses and have some of the characteristics of the endospores of some Gram positive bacteria (Kolter *et al.*, 1993; Loewen and Hengge-Aronis, 1994).

The survival response has two main objectives:
1. to ensure survival of the stress.
2. to prime the bacteria for growth when the stress is removed.

2 STRATEGIES FOR SURVIVAL

2.1 The Numbers Game

The amplification phase of *E. coli* in the intestines of warm blooded animals and birds produces enormous populations of bacteria and this, together with a dispersal mechanism linked to animal defecation, ensures that huge numbers of *E. coli* are released into the environment, thereby maximising the chance that at least some of them will survive to find another host.

2.2 Environmental Reservoirs

When nutrients and conditions permit *E. coli* is able to multiply in a variety of habitats in the environment (Roszak and Colwell, 1987), e.g., biofilms (Block *et al.*, 1996; Jones and Bradshaw, 1996) and lake water (Ashbolt *et al.*, 1996) and is concentrated by filter-feeding shellfish (Lees, 1996). These habitats consequently serve as environmental reservoirs from which animals can be infected. Other coliforms such as *Klebsiella pneumoniae*, *Enterobacter agglomerans* and *Citrobacter freundii*, with their potential for nitrogen fixation, are better adapted for growth in the environment.

2.3 Shelter

E. coli and coliforms can be protected from unfavourable stresses by sheltering from them. For example, shade from a floating mat of *Lemna gibba* L provided protection from the biocidal effects of high intensity sunlight in a reservoir (Davedar and Bahgat, 1995); *Klebsiella pneumoniae* from sewage could not be isolated from the water column in Mediterranean and Pacific coastal waters but was isolated from sediments and biofilms on the underside of sea grasses and algae in the same habitat (Jones, 1991); and Perez-Rosas and Hazen (1988) demonstrated long term survival of *E. coli* in sand and amongst sea grasses on tropical coral reefs.

E. coli is 2,400 times more resistant to chlorine when attached to a surface than when free in the water column. Indeed, attachment of bacteria to surfaces is the main way that they survive disinfection in water delivery systems (LeChevallier *et al.*, 1988). Suspended particles in water provide bacteria with a supply of nutrients (they are nutritionally richer than the surrounding water column because of adsorbed organic matter) as well as shelter from irradiation and predation. *E. coli* released in sewage effluent is less susceptible to chlorination when attached to particles than when in the water column (Berman *et al*,. 1988). *E. coli* are also protected from chlorine when they are ingested, but not lysed, by protozoa (King *et al*., 1988).

2.4 Stress Survival

When *E. coli* and the other coliforms are released into the environment they are subject to a wide variety of stresses, e.g. osmotic shock, temperature shock and nutrient limitation. The ability of *E. coli* to survive stress depends on its position in the growth cycle. Bacteria from log phase are more sensitive to stress than those which have been through stationary phase or are in the viable but nonculturable phase (VBNC) (Figure 15.1) (Roszak and Colwell, 1987), e.g. acid tolerance of *E. coli* 0157:H7 in stationary-starvation phase is three times higher than in log phase (Benjamin and Datta, 1995).

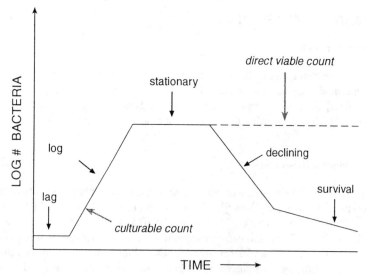

Figure 15.1 *Stages of the bacterial growth curve*

2.4a Stationary-Starvation-Survival

2.4a (i) Molecular/Genetic Control. During normal growth *E. coli* is able to detect nutrient depletion and enters stationary phase, a period of physiological and morphological adjustment. This is tightly controlled by the induction of genes which sequentially convert the actively growing cell into a resistant form. When *E. coli* becomes nutrient-stressed or enters stationary phase a nucleotide, guanosine 3',5'-bispyrophosphate (ppGpp), is induced (Spira *et al.*, 1995). This is a signal for the stringent response which is a co-ordinated shutting down of normal metabolic activities. Transcriptional control of RNA polymerase is switched from sigma factor σ^{70}, the housekeeping promoter responsible for controlling protein synthesis during normal growth, to σ^S, an alternate starvation sigma factor. σ^S directs the transcription of a series of overlapping networks of genes responsible for the production of a range of stress proteins which confer on *E. coli* resistance to a multiplicity of stresses (Arnold and Kaspar, 1995; Kolter *et al.*, 1993; Loewen and Hengge-Aronis, 1994; Nystrom, 1993 and 1995). A variety of other molecular controls are also activated but their functions are less clear (Nystrom, 1995).

When starvation is over Fis, a small DNA-binding protein, is induced, growth resumes and multiple stress resistance is lost.

2.4a (ii) Metabolic Changes during Starvation-Stationary-Survival. On entering stationary phase there is a flurry of protein synthesis lasting several hours. This coincides with the switch from normal metabolism to survival metabolism and the production of stress-related proteins. Protein synthesis thereafter drops rapidly and by 11 days is only 0.05% of a growing culture. Dps, a DNA binding protein which helps to offset oxidative stress, continues to be synthesised. Protein synthesis plays an important part in survival during starvation. If it is inhibited at the onset of a stress cells lose the ability to survive. Survival of *E. coli* is enhanced when nitrogen compounds, such as ammonium sulphate and amino acids, are present, perhaps because they stimulate protein synthesis. Stationary phase *E. coli* retain the ability to rapidly resume metabolism by retaining a functional electron transport chain and active membrane-bound ATPases (Siegele *et al.*, 1993). The proportion of cells staying viable depends on their condition at starvation and the type of starvation. Cryptic growth can occur when lysis of dead cells produces nutrients for others. Addition of glucose rapidly removes the ability of the cells to resist heat and other stresses.

In general starvation of nutrients other than carbon, e.g., nitrogen and phosphorus result in faster entry into the stationary/starvation phase, e.g., amino acid deprivation leads to a rapid stringent response and shutdown of growth (Rodionov and Ishiguro, 1995).

2.4a (iii) Morphological changes to the cell. Stationary-phase, VBNC and stressed *E. coli* cells are smaller and rounder than normal (Roszak and Colwell, 1987; Kolter *et al.*, 1993). This is the result of successive divisions, orchestrated by σ^S, with no increase in cell mass (Rockabrand *et al.*, 1995). Cytoplasm is condensed and the volume of periplasm increases. The cell envelope becomes covered with hydrophobic molecules which promote adhesion and aggregation, and some starved *E. coli* produce curli fibres which cause bacteria to clump together and to adhere to host cells by binding to fibronectin and laminin. This too is also under the control of σ^S (Loewen and Hengge-Aronis, 1994). Cell wall synthesis ceases (Kolter *et al.*, 1993) but the cell wall thickens due to an increase of the peptidoglycan layer from 2-3 layers to 4-5 layers and becomes less susceptible to autolysis. Changes in osmoregulation occur when new membrane synthesis ceases and membranes become less fluid and less permeable due to fewer porins and changes in fatty acids (Kolter *et al.*, 1993). In *K. pneumoniae* capsular polysaccharide is increased during nitrogen starvation and colonic acid, which binds many times its weight of water and protects colonies and individual cells against desiccation, builds up in the capsule (Ophir and Gutnick 1994).

2.4a (iv) Summary of Stationary-Starvation-Survival Resistance. After passing through stationary phase *E. coli* are resistant to multiple stresses, e.g.,

> nutrient starvation
> shifts in temperature
> acid
> high osmotic pressure and salinity
> UV radiation
> oxidative stress
> uptake of antibiotics

They are often more resistant to a stress than a growing cell which has been newly exposed to a stress and then becomes resistant. Although in stasis, they are metabolically active and immediately respond to the addition of nutrients or the removal of the stress. On growth they lose the multiple stress resistance acquired during stationary-starvation phase.

2.4b Survival of Individual Stresses

Each environmental stress activates a set of genes, a stimulon, specific to that stress (Kolter *et al.*, 1993). Their function is to repair damage caused by sub-lethal stress and, if the stress is severe, to activate the stringent response (as in stationary-survival), i.e., to shut down the cell in preparation for prolonged survival. As *E. coli* produces the same signal, ppGpp, in response to a variety of stresses and because the alternative regulator, σ^s, is the same for different stimulons, more than one set of gene products may be produced as a result of a stress. This molecular overlap is the reason for the observation that exposure of *E. coli* to one stress will frequently make it resistant to another, e.g., *E. coli* acquiring resistance to acid shock also becomes resistant to oxidative shock.

2.4b (i) Heat Shock. When exposed to a higher temperature *E. coli* quickly produces heat-shock proteins (hsps) to alleviate damage to proteins (Yura *et al.*, 1993). Hsps are molecular chaperones which assist unfolded, misfolded and aggregated proteins to regain shape and function and help translocation of proteins across membranes. They prevent inactivation of proteins, reactivate inactivated proteins and degrade non-repairable proteins that accumulate in stress conditions. Hsps, e.g., DnaK, GroES and GroEL, are overproduced at temperatures above 40°C and are important in protection against thermal stress.

Lelivelt and Kawula (1995) have shown that after a shift from 37°C to 10°C, *E. coli* produces a cold shock protein, Hsp66, which functions as a stress protein. PpGpp is the signal for the induction of the cold shock stimulon and, as it is produced by the ribosome associated RelA protein, they propose that the ribosome is a sensor for cold shock.

Hsps are also induced by many other stresses, e.g., DNA damaging agents, amino acid analogues, stringent response, starvation of C, bacteriophage, unfolded proteins, alkaline shift, and ethanol. They can, therefore, be regarded as general antistress proteins (Farr and Kogoma, 1991).

Heat shock has been shown to reverse the VBNC state in *E. coli* in drinking water (Berry *et al.*, 1991). A three log increase in the culturable count was obtained after heat treatment for only 20 minutes at 35°C, implying that simply ingesting water contaminated by VBNC would revive them.

2.4b (ii) Osmotic Shock. E.coli can detect and move away from unfavourable concentrations of osmolytes such as sodium chloride (Csonka, 1989). Motility and chemosensory signal processing in *E. coli* is regulated along with osmoregulation so that flagella assembly and movement are synchronised (Shin and Park, 1995). *E. coli* uses osmoregulation to maintain a constant cell volume over a range of external osmolarities. High external concentrations of salts induce a response controlled by σ^s. There are two main responses to osmotic shock: firstly, the *ompr* gene reduces porin proteins in the membrane to control uptake of osmolytes into the cell; and secondly, there is the uptake of non-toxic compatible solutes (Csonka, 1989). Several compatible solutes have been identified for *E. coli*: K^+, betaine, glutamate, glutathione, glycine-betaine, proline and trehalose, with glycine-betaine the preferred osmoprotectant. They accumulate in high concentrations to prevent growth inhibition. Compatible solutes are taken up in the following order: K^+ followed by glutamate; K^+ is then excreted and glutamate converted to glutamine. This is followed by trehalose production which replaces both K^+ and glutamate. If glycine betaine or proline is available K^+ is excreted and trehalose depleted. Expression of genes for K^+ uptake results directly from the increase in extracellular osmotic pressure but expression of betaine transport is induced by the increase in ionic strength following the primary uptake of K^+ (Wiggins, 1990). Control of proline and glycine-betaine uptake from stationary phase is by *proP* a membrane-bound ion transporter. *ProP* is regulated by two promoters, Fis and os. Unlike other Fis activated genes *proP* is maximally expressed in stationary phase (Xu and Johnson, 1995). Trehalose is both an osmo- and thermoprotectant which has the capacity to protect both proteins and membranes.

Although habituation to salinity can occur (Munro *et al*, 1987) sea water generally slows down and modifies growth in *E. coli* and induces the VBNC state (Anderson *et al.*, 1979; Roszak and Colwell, 1987; Gourmelon *et al.*, 1994). Survival of *E. coli* and other coliforms is enhanced in sea water and sediments by the presence of osmoprotectants such as glycine-betaine and proline, excreted by cyanobacteria, algae and higher plants (Munro *et al.*, 1989). Glycine-betaine, and proline-betaine in urine have been shown to protect pathogenic E.coli from hypertonic NaCl in kidneys (Chambers and Kunin, 1987).

2.4b (iii) Acid Shock. When growing in rich media *E. coli* produces decarboxylases to neutralise acid and deaminases to neutralise alkalinity (Slonczewski, 1992). The ability to withstand acid is important for the survival and pathogenicity of *E. coli* as it is exposed to low pH in macrophages, the stomach (Benjamin and Datta, 1995), faeces, farm slurries (Kearney *et al.*, 1993); and water (Wortman and Bissonnette, 1988). *E. coli* 0157:H7 has been isolated from low pH cottage cheese, cider, yoghurt and mayonnaise (Arnold and Kasper, 1995) and has been found to be resistant to mineral and organic acid washes used during the cleaning of animal carcasses, especially at low temperatures (Conner and Kotrola, 1995). The ability to withstand low pH is one of the reasons for *E. coli*'s dominance in a variety of cheeses (Tornaddijo *et al.*, 1993). Benjamin and Datta (1995) have shown that some strains of *E. coli* 0157:H7 can withstand 5 hours at the pH of gastric juices without loss of activity. Lin *et al.* (1995) found three responses to acid: acid habituation, acid tolerance response (ATR) and acid resistance (AR). ATR was induced by shifting log and stationary-phase cells to pH 4.3 (acid shock) which enabled the cells to resist further challenge with pH 3. AR was demonstrated only with stationary-phase cells which were exposed to pH 5 and became protected for survival at pH 2 and below. Acid habituation, followed brief exposures of log phase cells to pH 5, permitted survival at previously lethal doses. Three AR mechanisms were found: a low-pH-induced system expressed during aerobic growth and two others expressed during fermentation, one required arginine and the other glutamate. The amino acids may serve as precursors for an acid stress protein. Arnold and Kasper (1995) also found that stationary-starvation phase *E. coli* 0157:H7 were more acid resistant than those habituated in log phase and that there are two pH-dependent systems, one induced in log phase and the other in stationary phase. They also found a third which is pH independent and induced in stationary phase and involves the stationary-phase-associated sigma factor σ^s. Wortman and Bissonnette (1988) showed that less serious acid injury due to organic acids required only protein synthesis for repair while severe acid injury required RNA synthesis, protein synthesis and an energised membrane.

Acid resistance can be broken by exposure of E.coli to NaCl or alkaline pHs (Rowbury *et al.*, 1993). Mendonca *et al* (1994) have demonstrated that *E. coli* membranes are disrupted at high pH with no recovery and suggest that high pH treatments may be a better way of destroying food-borne Gram negative bacteria than acid.

2.4b (iv) Sunlight. Sunlight is such a powerful bactericide that Acra *et al.* (1989) suggest that it is used to disinfect drinking water and oral rehydration fluids in the tropics. There is a wealth of literature showing that sunlight is the major factor affecting the survival of *E. coli* in surface waters (e.g., Barcina *et al.*, 1989; Chan and Killick, 1995; Curtis *et al*, 1992; Davies and Evison, 1991; Davies-Colley *et al* 1994, Fujioka *et al*,1981; Gourmelon *et al*, 1994; Jones 1994; Sinton *et al.*,1994; Solic and Krstulovic 1992). T90 data, adapted from Wallace (1994), show that the affects of sunlight vary with the seasonal changes in light (Figure 15.2). In the winter months T90s were not reached, even after a full day's sunshine. The higher numbers seen in MPN compared to plate counts are due to dark repair during enrichment culture. Direct viable counts showed that the cells had entered the VBNC stage, a phenomenon seen by others (Barcina *et al.*, 1989; Curtis *et al.*,1992; Davies and Evison, 1991; Davies-Colley *et al.*,1994, Gourmelon *et al.*, 1994; McKay 1992; Roszak and Colwell, 1987; Solic and Krstulovic 1992). Using narrow bands of light, Wallace (1994) showed that UVB is the main cause of cell damage to *E. coli* but other workers have suggested that UVA

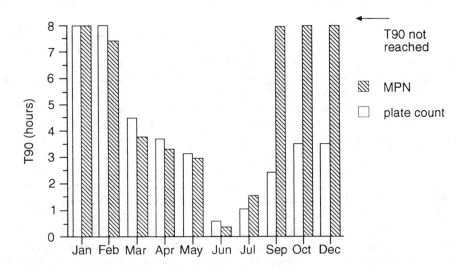

Figure 15.2 *Seasonal effects of sunlight on T90s for Escherichia coli*

and visible light are more important, largely due to lack of penetration of UVB into natural waters (Barcina *et al.*, 1989; Curtis *et al.*, 1992; Davies-Colley *et al.*, 1994, Gourmelon *et al.*, 1994; Sinton *et al.*, 1994). The different wavelengths of light, shown below, damage cells in different ways:

UVC	(far UV)	200-280nm
UVB	(mid UV)	280-320nm
UVA	(near UV)	320-400nm
Visible		400-700nm

UVC is adsorbed by nucleic acids resulting in the formation of pyridine dimers.
UVA and visible light cause widespread damage to membranes, DNA, RNA and enzymes.
   It requires aerobic metabolism and photosensitising molecules.
UVB causes both types of damage.
UVC, which is becoming widely used in the water industry, is 100 times more effective
   when modulated rather than continuous (Bank *et al.*, 1990).

Sub-lethal damage to nucleic acids caused by UVC and UVB is repaired by two main processes, one requiring visible light, photoreactivation, and the other carried out in the dark. Dark repair is subdivided into, excision, recombinational and SOS (error prone), with excision the most important. UVA and visible light cause the formation of active oxygen species which induce the oxidative stress response (Goodson and Rowbury, 1991; Harris *et al.*, 1987a and 1987b; Jagger *et al.*, 1964; Kelland *et al.*, 1983; Sammartano *et al.*, 1986; Schobert and Kolch, 1992; Tuveson *et al.*, 1988; Yammamoto *et al.*, 1983).

   2.4b (v) Oxidative Stress. Oxygen toxicity occurs when the degree of oxidative stress exceeds the capacity of the defence systems (Farr and Kogoma, 1991). Active oxygen molecules damage DNA, RNA, protein and lipids and are formed during normal aerobic metabolism. Their formation is enhanced during exposure to other stresses such as light and heat. Active oxygen species include: superoxide radical $[O_2-]$, hydrogen peroxide $[H_2O_2]$, hydroxyl radical $[OH]$, singlet oxygen $[1\Sigma O_2]$, hydroperoxyl radical $[HOO]$ and ozone and are produced in a variety of ways (see Farr and Kogoma, 1991).

   When exposed to active oxygen there is a two staged response: to eradicate the active oxygen species and to repair damage to DNA and membranes. *E. coli* has superoxidedismutases (SOD) which convert O_2- to H_2O_2 and catalases which convert H_2O_2 to H_2O and O_2. When these processes cannot cope oxidative stress occurs and at least two oxidative stress stimulons are induced:

1. The peroxide stress stimulon, which is regulated by the *OxyR* regulon. DNA is repaired
   and cells show increased resistance to H_2O_2.
2. The superoxide stress stimulon, which is regulated by *sox R* and *sox S*, and provides
   increased DNA repair capacity when stressed by O_2-.

Non enzyme antioxidants, such as glutathione, carotenoids, thioredoxin, ubiquinone and menaquinone maintain a reduced cellular environment, and glutathione reduces the sulphide bridges in proteins caused by oxidative stress. Severe oxidative stress will induce stringent, heat shock and osmotic shock responses.

   Oxidative stress proteins are formed during starvation-stationary phase in *E. coli* but, unlike those induced by direct oxidative stress, their production is controlled by σ^S.

2.4b (vi) Stress Response Overlap. Exposure to one stress will often convey resistance to another. Acid habituation causes resistance to UV, oxidative stress resistance to heat shock and damage to DNA, and starvation resistance to heat, oxidative and osmotic shocks. This is because of overlap at the regulatory level, e.g., the hsps (DnaK, GroEl, HtpG, regulated by σ^s) are induced by peroxide, superoxide, cold and heat shock, starvation, DNA damage and antibiotics such as naladixic acid.

Whilst there is overlap in the response to stresses there is also synergy in the effect of stresses, e.g., light induced entry into the VBNC state is more rapid in salt rather than fresh water (Gourmelon *et al.*, 1994) and the rates of photoreactive and dark repair of UV damage are altered by salinity, light and temperature (Chan and Killick, 1995).

2.5. Viable but nonculturable cells

The ability to form VBNC cells is recognised as an important strategy in the survival of *E. coli* and the other coliform bacteria and has important implications for water and food microbiology (Roszak and Colwell, 1987). VBNC cells are metabolically active, but incapable of cell division on media which normally support their growth (Oliver, 1993). They have the survival characteristics of stationary-starvation cells but will not grow unless ingested by a suitable host or their dormancy is otherwise reversed. VBNC *E. coli* have been found to form in drinking water (Byrd and Colwell, 1991) and their dormancy reversed by relatively mild heat shock (Berry *et al.*, 1991). The longevity and infectivity of VBNC *E. coli* are matters for conjecture.

3 SUMMARY

The ability of *E. coli* to become resistant to a multiplicity of stresses on entering stationary-starvation phase or on experiencing a stress has implications for the way in which we interpret survival. For example, *E. coli* in sewage effluent and discharged into receiving waters will be in a starved, and therefore resistant state. It will survive for longer than tests done on log phase bacteria would predict. The capacity for repair after a stress shows why regrowth can occur some time after treatment, e.g., photoreactivation after UV treatment. Survival in high concentrations of salts and sugars or at low pHs in food may be the result of resistance acquired during stationary-starvation or the bacteria becoming habituated or resistant after exposure to an unrelated stress. For the microbiologist the problem of *E. coli*'s survival is confounded by the formation of the VBNC state which can not be detected by the usual cultural methods.

References

Acra A, Jurdi M, Mu Allem H, Karahagopian Y, Raffoul Z. Sunlight as disinfectant. Lancet 1989;333:280.

Anderson IC, Rhodes MW, Kator HI. Sub lethal stress in *Escherichia coli*, a function of salinity. Appl. Environ. Microbiol. 1979;38:1147-1152.

Arnold KW, Kaspar CW. Starvation- and stationary-phase induced acid tolerance in *Escherichia coli* 0157:H7. Appl. Environ. Microbiol. 1995;61:2937-2039.

Ashbolt NJ, Dorsch M, Banens B. Blooming *E. coli*. What do they mean. This volume, 1996.

Bank HL, John J, Schmehl MK, Dratch RJ. Bactericidal effectiveness of modulated UV light.

Appl. Environ. Microbiol. 1990;56:3888-3889.

Barcina I, Gonzalez JM, Iriberri J, Egea L. Effect of visible light on progressive dormancy of *Escherichia coli* cells during the survival process in natural freshwaters. App. Environ. Microbiol. 1989;55:246-251.

Berman D, Rice EW, Hoff JC. Inactivation of particle -associated coliforms by chlorine and monochloramine. Appl. Environ. Microbiol. 1988;54:507-512.

Benjamin MM, Datta AR. Acid tolerance of enterohemorrhagic *Escherichia coli*. Appl. Environ. Microbiol. 1995;61:1669-1672.

Berry C, Lloyd BL, Colbourne JS. Effect of heat shock on recovery of *Escherichia coli* from drinking water. Wat. Sc. Tech. 1991; 24:85-88.

Block JC, Mouteaux L, Gatel D, Reasoner DJ. Survival and growth in a drinking water distribution system. This volume. 1996.

Byrd JJ, Colwell RR. Viable but nonculturable bacteria in drinking water. Appl. Environ. Microbiol. 1991;57:875-878.

Chan YY, Killick EG. The effect of salinity, light and temperature in a disposal environment on the recovery of *Escherichia coli* following exposure to ultraviolet radiation. Wat. Res. 1995;29:1373-1377.

Conner DE, Kotrola JS. Growth and survival of *Escherichia coli* 0157:H7 under acid conditions. Appl. Environ. Microbiol. 1995;61:382-385.

Csonka LN. Physiological and genetic responses of bacteria to osmotic stress. Microbiol. Reviews 1989;53:121-147.

Curtis TP, Mara DD, Silva SA. Influence of pH, oxygen, and humic substances on the ability of sunlight to damage faecal coliforms in waste stabilisation pond water. Appl. Environ. Microbiol. 1992;58:1335-1343.

Davedar A, Bahgat M. Fate of faecal bacteria in a waste water retention reservoir containing *Lemna gibba* L. Wat. Res. 1995;29:2598-2600.

Davies CM, Evison LM. Sunlight and the survival of enteric bacteria in natural waters. J. Appl. Bacteriol. 1991;70:265-274.

Davies-Colley RJ, Bell RG, Donnison AM. Sunlight inactivation of enterococci and faecal coliforms in sewage effluent diluted in sea water. Appl. Environ. Microbiol. 1994;60:2049-2058.

Farr SB, Kogoma T. Oxidative stress responses in *Escherichia coli* and *Salmonella typhimurium*. Microbiol. Rev. 1991;55:561-585.

Fujioka RS, Hashimoto HH, Siwak EB, Young RHT. Effect of sunlight on survival of indicator bacteria in sea water. Appl. Environ. Microbiol. 1981;41:690-696.

Goodson M, Rowbury RJ. Rec A. Independent resistance to irradiation with UV light in acid-habituated *Escherichia coli*. J. Appl. Bacteriol. 1991;70:177-180.

Gourmelon M, Cillard J, Pommepuy M. Visible light damage to *Escherichia coli* in sea water: oxidative stress hypothesis. J. Appl. Bacteriol. 1994;77:105-112.

Harris GD, Adams VD, Sorensen DL, Curtis MS. Ultraviolet inactivation of selected bacteria and viruses with photoreactivation of the bacteria. Wat. Res. 1987a;21:687-692.

Harris GD, Adams VD, Sorensen DL, Dupont RR. The influence of photoreactivation and water quality on ultraviolet disinfection of secondary municipal wastes. J. Wat. Poll. Con. Fed. 1987b;59:781-787.

Hengge-Aronis R. The role of rpoS in early stationary-phase gene regulation *Escherichia coli* K12. In Kjelleberg S, editor. Starvation in Bacteria. London: Plenum, 1993:171-200.

Jagger J, Wise WC, Stafford RS. Delay in growth and division by near-ultraviolet radiation in

Escherichia coli B and its role in photoprotection and liquid holding recovery. Photochem. Photobiol. 1964;3:11-24.

Jones K. Waterborne diseases. New Scientist Inside Science 73, 1994:1-4.

Jones K. A comparison of the distribution of heterotrophic nitrogen-fixing bacteria in coastal waters of Morecambe Bay, UK, the Ligurian Sea, France, the Bay of Naples, Italy, and the Pacific Ocean, Hawaii, USA. In Elliot M, Ducrotoy J-P, editors. Estuaries and Coasts: Spatial and Temporal Intercomparisons. Fredensborg: Olsen and Olsen, 1991:111-116.

Jones K, Bradshaw S. Biofilm formation by the Enterobacteriaceae: a comparison between *Salmonella enteritidis*, *Escherichia coli*, and a nitrogen-fixing strain of *Klebsiella pneumoniae*. J. Appl. Bacteriol. 1996; in press.

Kapuscinski RB, Mitchell R. Solar radiation induces sub lethal injury in *Escherichia coli* in sea water. Appl. Environ. Microbiol. 1981;41:670-674.

Kearney TE, Larkin MJ, Levett PN. The effect of slurry storage and anaerobic digestion on survival of pathogenic bacteria. J. Appl. Bacteriol. 1993;74:86-93.

Kelland LR, Moss SH, Davies DJG. Recovery of *Escherichia coli* K-12 from near-ultraviolet radiation-induced membrane damage. Photochem. Photobiol. 1983;37:617-622.

King CH, Shotts EB, Wooley RE, Porter KG. Survival of coliforms and bacterial pathogens within protozoa during chlorination. Appl. Environ. Microbiol. 1988;54:3023-3033.

Kolter R, Siegele DA, Tormo A. The stationary phase of the bacterial life cycle. Ann. Rev. Microbiol. 1993;47:855-874.

LeChevallier MW, Singh A, Schiemann DA, McFeters GA. Changes in virulence of waterborne enteropathogens with chlorine injury. Appl. Environ. Microbiol. 1984;50:412-419.

LeChevallier MW, Cawthon CD, Lee RD. Inactivation of biofilm bacteria. Appl. Environ. Microbiol. 1988;54:2492-2499.

Lees DN. Faecal coliforms in shellfish. This volume, 1996.

Lelivelt MJ, Kawula TH. Hsc66, an Hsp70 homolog in *Escherichia coli*, is induced by cold shock but not by heat shock. J. Bacteriol. 1995;177:4900-4907.

Lin J, Lee IS, Frey J, Slonczewski JL, Foster JW. Comparative analysis of extreme acid survival in *Salmonella typhimurium*, *Shigella flexneri*, and *Escherichia coli*. J. Bacteriol. 1995;177:4097-4104.

Loewen PC, Hengge-Aronis R. The role of the sigma factor σ^s (KatF) in bacterial global regulation. Ann. Rev. Microbiol. 1994;48:53-80.

McKay AM. Viable but non-culturable forms of potentially pathogenic bacteria in water. Lett. Appl. Microbiol. 1992;14:129-135.

Mendonca AF, Amoroso TL, Knabel SJ. Destruction of gram negative food-borne pathogens by high pH involves disruption of the cytoplasmic membrane. Appl. Environ. Microbiol. 1994;60:4009-4014.

Munro PM, Laumond F, Gautier MJ. A previous growth of enteric bacteria on a salted medium increases their survival in sea water. Lett. Appl. Microbiol. 1987;4:121-124.

Munro PM, Gautier MJ, Breittmayer VA, Bongiovanni J. Influence of osmoregulation processes on starvation survival of *Escherichia coli* in sea water. Appl. Environ. Microbiol. 1989;55:2017-2024.

Nystrom T. Global systems approach to the physiology of the starved cell. In Kjelleberg S, editor. Starvation in Bacteria. London: Plenum, 1993:129-150.

Nystrom T. The trials and tribulations of growth arrest. Trends Microbiol. 1995;3:131-136.

Oliver JD. Formation of viable but non-culturable cells. In Kjelleberg S, editor. Starvation in Bacteria. London: Plenum, 1993:239-272.

Ophir T, Gutnick DL. A role for exopolysaccaride in the protection of microorganisms from desiccation. Appl. Environ. Microbiol. 1994;60:740-745.

Perez-Rosas N, Hazen TC. In situ survival of *Vibrio cholerae* and *Escherichia coli* in tropical coral reefs. Appl. Environ. Microbiol. 1988;54:1-9.

Rockabrand D, Arthur T, Korinek G, Lives K, Blum P. An essential role for the *Escherichia coli* DnaK protein in starvation-induced thermotolerance, H_2O_2 resistance and reductive division. J. Bacteriol. 1995;177:3695-3703.

Rodionov DG, Ishiguro EE. Direct correlation between over production of guanosine 3',5' bispyrophosphate (ppGpp) and penicillin tolerance in *Escherichia coli*. J. Bacteriol. 1995;177:4224-4229.

Roszak DB, Colwell RR. Survival strategies of bacteria in the natural environment. Microbiol. Rev. 1987;51:365-379.

Rowbury RJ, Goodson M. PhoE porin of *Escherichia coli* and phosphate reversal of acid damage and killing and of acid induction of the CadA gene product. J. Appl. Bacteriol. 1993;74:652-661.

Sammartano LJ, Tuveson RW, Davenport P. Control of sensitivity to inactivation by H_2O_2 and broad spectrum near-UV radiation by the *Escherichia coli* KatF locus. J. Bacteriol. 1986;168:13-21.

Schobert B, Kolch A. Photoreactivation of *Escherichia coli* depending on light intensity after UV radiation. Zentral. Hyg. Umweltmed. 1992;192:565-570

Shin S, Park C. Modulation of flagella expression in *Escherichia coli* by acetyl phosphate and the osmoregulator Ompr. J. Bacteriol. 1995;177:4696-4702.

Siegele DA, Almiron M, Kolter R. Approaches to the study of survival and death in stationary-phase *Escherichia coli*. In Kjelleberg S, editor. Starvation in Bacteria. London: Plenum, 1993:151-169.

Sinton LW, Davies-Colley RJ, Bell RG. Inactivation of enterococci and faecal coliforms from sewage and meat works effluents in sea water chambers. Appl. Environ. Microbiol. 1994;60:2040-2048.

Slonczewski JL. pH-regulated genes in enteric bacteria. ASM News 1992;58:140-144.

Solic M, Krstulovic N. Separate and combined effects of solar radiation, temperature, salinity and pH on the survival of faecal coliforms in sea water. Mar. Poll. Bull. 1992;24:411-416.

Spira B, Silverstein N, Eggeling L. Guanosine. 3',5'-biphosphate (ppGpp) synthesis in cells of *Escherichia coli* starved for Pi. J. Bact. 1995;177:4053-4058.

Tornaddijo E, Fresno JM, Carballo J, Martin-Sarmiento R. Study of Enterobacteriaceae throughout the manufacturing and ripening of hard goats' cheese. Appl. Bacteriol. 1993;75:240-246.

Tuveson RW, Larson RA, Kagan J. Role of cloned carotenoid genes expressed in *Escherichia coli* in protection against inactivation by near-UV light and specific phototoxic molecules. J. Bacteriol. 1988;170:4675-4680.

Wallace JS. The effects of light and other abiotic factors on the survival of the indicator bacteria *Escherichia coli* and *Klebsiella pneumoniae* and the pathogenic bacteria *Campylobacter jejuni* and *Salmonella* enteritidis. PhD thesis, Lancaster University, Lancaster, 1994.

Wiggins PM. Role of water in some biological processes. Microbiol. Rev. 1990;54:432-449.

Wortman AT, Bissonnette GK. Metabolic processes involved in repair of *Escherichia coli* cells damaged by exposure to acid mine water. Appl. Environ. Microbiol. 1988;54:1901-1906

Xu J, Johnson RC. Fis activates the RpoS-dependent stationary-phase expression of proP in *Escherichia coli*. J. Bacteriol. 1995;177:5222-5231.

Yammamoto KM, Satake M, Shinagawa H, Fujiwara Y. Amelioration of the ultraviolet sensitivity of *Escherichia coli* recA mutant in the dark by photoreactivating enzyme. Molec. Gen. Genet. 1983;190:511-515.

Yura T, Nagai H, Mori H. Regulation of the heat shock response in bacteria. Ann. Rev. Microbiol. 1993;47:321-350.

Chapter 16
Bacteria In Recreational Waters: A Regulator's Concerns

Paul R. Holmes

Environmental Protection Department
Government of Hong Kong

1 THE REGULATOR'S PROBLEM

The pollution control authority has a responsibility to ensure that pollution controls are adequate and effective. Monitoring effectiveness is especially important when, as in Hong Kong, the one authority is responsible for both the provision of community water pollution control works and the operation of the pollution control laws.

For recreational waters, the regulator's task may resolve to two distinct questions: one when looking outward to the recreating public, and the other inward to the professionals responsible for sewage disposal. The outward question is the simpler to formulate and the harder to answer: "Is the water safe for recreation?" The inward question asks what standards must be met in the treatment and disposal of sewage and waste water, including the treatment required and the location of outfalls in relation to recreational areas. Increasingly, such areas are not just bathing beaches but include much remoter locations accessible to windsurfers, dinghy sailors, surf canoeists, other boat users and scuba divers.

It is many years since a pollution control authority could declare with impunity that swimming in sewage-contaminated water would not cause significant health risk unless the pollution were so gross as to be "aesthetically revolting". Today's regulator must respond to stricter community goals.

Controversy over standards for recreational water quality has continued for many years. Faecal coliform counts have been used as indicators of sewage pollution, and by implication of the risk of pathogens being present, for over 80 years (Hodgkiss, 1989). As Hodgkiss points out, there remains great diversity in the standards promoted or enforced by different authorities and experts. Research over the past two decades has led not to consensus, but to the proposal of numerous alternative standards, so deepening the regulator's confusion.

Several researchers have applied "imported" standards to their field areas, often in the process drawing attention to the deficiencies of the standards. For example, Kay (1988) applied current European and North American standards to Welsh beaches, while questioning the validity of *Escherichia coli* as the indicator of faecal pollution. Hodgkiss (1989) applied European and World Health Organization standards to various waters in Hong Kong. He found that these standards were frequently exceeded and questioned the validity of the then current Hong Kong standard for bathing waters.

Such reports may or may not reveal anything of importance about how effective the pollution control authorities are. They may provoke a polarization of attitudes between an environmentalist movement and a defensive pollution control body. Rationally, however, they are relevant to monitoring the performance of the pollution control authority only if the standards themselves are appropriate.

That internationally known standards are appropriate is by no means axiomatic. Federal water quality criteria in the USA, based on the epidemiological work of Cabelli and co-workers (Cabelli *et al.*, 1982; Cabelli, 1983) came under attack. Fleisher (1991) drew attention to "serious analytical and methodological weaknesses", suggesting that the basis of the criteria was too weak to justify changing from the traditional coliforms to enterococci, as proposed by Cabelli.

A modified epidemiological approach promoted in the UK, for example by Godfree *et al.* (1990), involved the exposure of healthy volunteers to bathing waters that met existing standards. Careful work over four seasons led to this team's proposing faecal streptococci concentrations, which showed a significant dose-response relationship with gastroenteritis, as a better indicator of water quality (Kay *et. al.*, 1994). However, Cabelli's method continued, with some variations, to be used and defended (Cheung and Chang , 1991, Wheeler and Alexander, 1992, Kueh *et al.*, 1995).

While both field evidence and controversy were accumulating, no one suggested that there was no risk to the health of bathers who exposed themselves to sewage-contaminated water. At least one expert, however, came to the aid of the now befuddled regulator. Pike (1993), pointing out that only improvements in sewage disposal arrangements could improve water quality, asked "whether standards are needed if there is a national policy for steadily improving marine treatment facilities for sewage".

As Pike noted, there is no consistency in the relationship between health risks, levels of pathogens and indicators, and none of the standards in use in Europe and North America are satisfactory. The population at large is not necessarily concerned, as superficial perceptions dominate the assessment of risk (Holmes, 1988).

So, the suggestion of abandoning the search for an ideal indicator in favour of getting on with the job of pollution control is an attractive one. Unfortunately, it does not lend a ready answer to either of the regulator's fundamental questions. Would such an approach satisfy an increasingly sophisticated and demanding public? Such a public, motivated by increasing prosperity and leisure time throughout the world's more economically developed countries, may expect unrealistically high environmental standards (Johnstone and Horan, 1994). Rationality does not increase with either sophistication or prosperity, and such expectations are more likely to be based on sensory perception than any scientific assessment of underlying risks (Holmes, 1988).

2 POLLUTION OF RECREATIONAL WATERS IN HONG KONG

Hong Kong is a territory comprising about 1000 km^2 of land and 1800 km^2 of sea, including many islands, situated on the edge of the Pearl River delta in southern China. Its long coastline, including both steep, rocky shores and broad sandy beaches, lends itself to water recreation. The density of population, with six million people living in just ten per cent of the land area, creates special pressure on those natural recreational resources that are close at hand.

On the other hand, Hong Kong's people and thriving business enterprises produce about two million tonnes of sewage and liquid industrial effluent every day. About half is untreated and 40 per cent receives only preliminary treatment before it is discharged into the sea. In terms of population equivalent, Hong Kong's pollution load on the sea is remarkably similar to direct discharges in the United Kingdom (Anderson, 1992), which has a land area over 240 times greater. The waters of Victoria Harbour, in the urban core of Hong Kong, have suffered steadily declining water quality over the two decades of monitoring. Water quality in the peripheral more scenic areas, including a number of popular bathing beaches, remains good to fair (Holmes and Smith-Evans, 1995).

During the 1970s and most of the 1980s, the Hong Kong government's water pollution control strategy was built around the provision of about 15 submarine sewage outfalls, with preliminary treatment only, in the main urban areas. Whilst new towns were provided with a higher standard of facilities, small communities in the hinterlands of bathing beaches and other scenic areas were left little or no communal provision. In the early years of these decades, this mattered little, but with increasing prosperity, the population increasingly migrated out of the urban core towards the more attractive living environment of scenic coastal areas. Septic tanks that had served an earlier generation were irrelevant.

Building law required that new housing developments be provided with some form of sewage treatment facilities, usually comprising a package plant. The law did not contain any enforceable provision to ensure that such plant be maintained in working order, or even that it be run at all, still less that it provide effective disinfection for its effluents. As a result, by the mid 1980s, some of the most popular bathing beaches, receiving as many as fifty thousand visitors on a good day, became heavily polluted.

A rather arbitrary standard for Hong Kong bathing beaches had been established in 1981. It required that the running median *E. coli* count of the five most recent samples taken at intervals of between three and fourteen days during the bathing season should not exceed 1000 dl^{-1}. By 1987, even this lax standard was frequently breached at some popular beaches, and a new "acceptability criterion" was adopted. The limit was set at sixty per cent compliance, during the bathing season, with an *E. coli* count of 1000 dl^{-1}. This degree of pollution was actually, if unknowingly, accepted by a large number of bathers at popular beaches. (Holmes, 1988). The pollution provoked increasing criticism (Hodgkiss, 1988; Morton, 1989) while researchers found worrying signs of the underlying risks that the indicator might represent, such as a high degree of antibiotic resistance in faecal bacteria in streams flowing through populated areas (French *et al.*, 1987).

The Environmental Protection Department, established in its present form in 1986, initially could do little but observe the problem (Holmes, 1994). The Water Pollution Control Ordinance could be enforced only within designated areas, and at the time it provided very generous exemptions from control for pre-existing effluents (Holmes, 1992). One very significant exception might be made: if it could be shown that an effluent caused a risk to health, the authority could cancel an exemption and bring the effluent under effective control. This provision, together with a desire to provoke the community into making informed choices in support of pollution control, led the Environmental Protection Department to undertake a prospective epidemiology study along the lines advocated by Cabelli *et. al.*,(1982).

3 EPIDEMIOLOGY STUDIES AT HONG KONG BEACHES

Following a small-scale orientation study in 1986, the main study was conducted at Hong Kong bathing beaches in the summer of 1987. Over thirty thousand beachgoers were contacted, and 18,741 usable responses were obtained. Exposure to bacteria in the bathing water was determined in terms of a number of microbiological indicators, averaged over each day of observation at any one location. (Cheung *et al.*,1991).

Several symptom groups were recorded, including "highly credible gastroenteritis" used by Cabelli to include vomiting and diarrhoea. The study showed that bathing in the coastal beaches of Hong Kong posed an increased risk of developing gastrointestinal, ear, eye, skin, respiratory symptoms. The total of all symptoms was also higher among swimmers. However, the number of swimmers reporting symptoms was less than in the American work.

In contrast to Cabelli's work in the USA, E. coli was found to be the best indicator for swimming- associated gastroenteritis and skin symptoms amongst the bathers. The occurrence of swimming- associated gastrointestinal, skin and respiratory symptoms, and the total of all symptoms, were significantly higher for swimmers in the more polluted waters. A linear correlation was established between E. coli counts and the sum of reported highly credible gastroenteritis and skin symptom rates (Figure 16.1).

The study also found that staphylococcus was a good indicator for ear, respiratory and total symptom counts. Indeed only staphylococcus showed a significant correlation with the total symptom counts. Enterococcus and faecal streptococcus did not show any correlation with any of the symptom groups in any combination.

This work showed that swimming in sewage-contaminated waters was a significant public health problem. Cheung et al. (op. cit) assumed that E. coli represented pollution by sewage, while staphylococci would represent pollution from bathers themselves. They estimated that, given the high rate of use of beaches in Hong Kong, there would be 60,800 excess cases of illness due to swimming in one year.

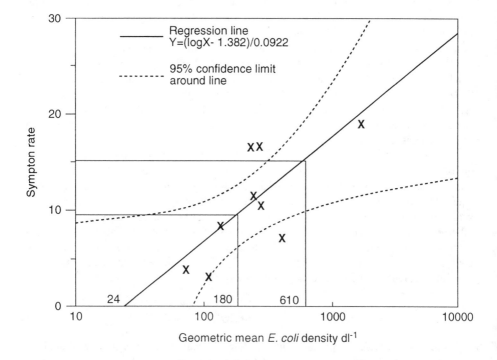

Figure 16.1 *Relationship between combined swimming-associated gastroenteritis and skin symptom rates and geometric mean E. coli densities at nine Hong Kong beaches in 1987 (redrawn from Cheung and Chang, 1988).*

New beach water quality objectives of 180 *E. coli* dl^{-1} and 1000 staphylococci dl^{-1} were proposed, both for beach acceptability and sewage disposal design criteria. A classification based on swimming-associated health risks was adopted: beaches are listed as **good**, with *E. coli* counts below 24 and no detectable excess illness; **fair**, *E. coli* from 25 to 180 and up to ten cases of gastrointestinal and skin symptoms expected per 1000 swimmers; **poor**, *E. coli* from 181 to 610 and up to 15 cases; and **very poor**, with *E. coli* counts over 610.

The significance of the cut-off at 610 *E. coli* dl^{-1} is that this was the observed degree of pollution at the most polluted popular bathing beach during the study, and was taken as the limit of acceptability. The beach management authorities would be expected to close beaches that did not consistently meet this standard, and have done so in a few cases. However, such decisions were taken reluctantly and only on an annual basis, considering a whole bathing season's results.

Meanwhile, the bacterial water quality and health risk levels of individual beaches are published every two weeks and in an annual summary report. In theory, this enables beach-goers to choose where to go for swimming based on current health risks, but our hope that information would make the annual decisions on beach closure redundant have not materialized. In particular, the public and the government's vocal critics in the environmental movement gave scant credence to the fact that at 15 gastrointestinal and skin cases per 1000 swimmers, the acceptability criterion compares favourably with the US federal standard's 19 gastrointestinal cases only. Murky water and the presence of plastic and other litter have told a far more convincing story than esoteric laboratory results, confirming again that perception is far more important than objective fact in assessing community response to risks.

For a time, the regulator had a rational means to assess the effectiveness of pollution controls for recreational waters in Hong Kong. The public could be informed about the risks swimmers faced, and the design engineers had design criteria. There were, of course, some problems with the epidemiology study: reliance on self diagnosis for reporting symptom rates and the dubious validity of the regression model, for example, but no better information was available. Most importantly, the findings did result in effective attention to the pollution problems. By 1990 water quality at many beaches, including some of the most polluted, had been greatly improved, although increasing pressures of urbanization have rolled back some of these gains subsequently (Figure 16.2).

There remained a nagging uncertainty about the exposure of bathers to the pathogens that might cause illness, exacerbated by the accumulation of evidence that pathogens are not conveniently related to indicator organisms. This led to a further epidemiological study in Hong Kong in 1992, involving another twenty-five thousand interviews and many more microbiological determinations (Kueh *et al.*, 1995). In an attempt to overcome some of the valid criticisms of earlier work, Kueh and her colleagues used laboratory testing for bacteria and viruses to complete the diagnosis of symptoms reported. They looked for a number of pathogens in the bathing water, using composite sampling to determine exposure.

There were serious differences between the results of this study and that carried out by Cheung *et al.*, (1991) in 1987. Kueh *et al.*, (*op. cit.*) reported the much higher figure of 41 cases of all symptom groups per 1000 swimmers, compared with only 30 in 1987. They found that *Clostridium perfringens* and *Aeromonas spp.* were significantly correlated with both gastrointestinal and highly credible gastrointestinal symptoms, but *E. coli* and faecal coliforms were not correlated with the occurrence of any symptoms. Other symptoms investigated and the total of all symptoms reported were not correlated with any indicators. Turbidity, however, was correlated with gastrointestinal symptoms. The authors ascribed this

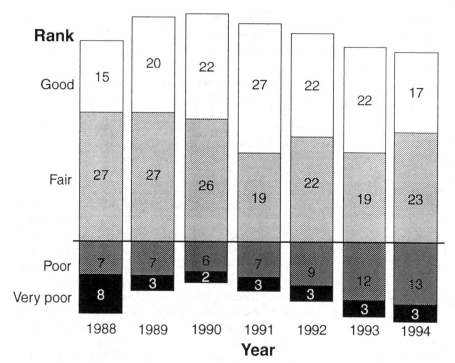

Figures indicate number of beaches in each rank in the stated year

Figure 16.2 *Ranks of Hong Kong bathing beaches from 1988 to 1994 (redrawn from Tam et al. 1994).*

to the effect of sewage discharges, although it may have been a matter of the perception of dirtiness resulting in a greater number of symptom reports by swimmers. Kueh *et al.* suggested that a turbidity standard of 15 NTU be added to the objectives, testing for *C. perfringens* and *Aeromonas spp.* being too hard for inclusion in a practical standard.

4 CRITIQUE OF THE METHODOLOGY

The divergence of results in the Hong Kong epidemiological studies reopened the regulator's questions. Taken together, these studies might be the most convincing field validation of Lacey and Pike's (1989) theoretical critique of the methodology developed by Cabelli and adopted in Hong Kong as elsewhere. Whilst acknowledging that the results of such studies support the conclusion that swimming increases the risk of minor ailments over not swimming, Lacey and Pike pointed out a number of discrepancies in the methodology and concluded that the results do not support a predictive relationship between indicator and illness.

These criticisms have been taken up and developed by a number of authors (for example Godfree, Jones and Kay, 1990; Fleisher, 1991; Fleisher *et al.*, 1993). In outline, the problems which some studies appear not to have addressed adequately are these:

- It is difficult to ensure that control and exposed cohorts are similar in all respects other than the exposure to bathing water; age banding may help but is an insufficient way to tackle this issue.
- Exposure of the individual bather to the indicator organisms tested is imprecisely known; bacteria densities vary greatly during the day and from place to place, even over a short distance.
- The exposure to pathogens implied by a given exposure to indicator organisms is not constant; there is a rapidly growing body of information showing that the survival of sewage-derived bacterial pathogens in the sea is very different from that of indicator organisms, and that the differential varies greatly with conditions of temperature, light, salinity, acclimatization, and other factors (for example Evison, 1988; Kueh and Grohmann, 1989; Pereira and Alcântara, 1993; Alkan *et al.*, 1995; Dupray and Derrien, 1995; Mezrioui *et al.*,1995; Wyer *et al.*,1995). Bacterial indicators may be totally irrelevant to viral pathogens.
- The linear regression model, although convenient and simple, is invalid for some of the data sets to which it has been applied. At very least, the independent variable, the exposure to water pollution, should be known precisely. In some cases, data from disparate sources have been pooled inappropriately.
- Some of the results have been extremely susceptible to data manipulation and over interpretation.
- Self-diagnosis of symptoms, especially when the subjects are aware of the nature of the investigation, inevitably poses a risk of over-reporting. Even clinical diagnosis and laboratory testing cannot completely overcome this problem, "feeling ill" being such a subjective matter.

A modification of the standard method involving controlled exposure of healthy volunteers to bathing water of precisely measured quality overcomes some of these difficulties (Kay *et al.*,1994). Using linear trend and multiple logistic regression techniques these authors concluded that only faecal streptococci showed a significant dose-response relationship with gastroenteritis. Unfortunately, as this method solves certain problems it creates or exaggerates others: the exposed population is limited to healthy adults, the water quality tested is restricted to the cleaner end of its natural range and, most importantly, the problem of symptom reporting and recognition is exaggerated by the subjects' inescapable awareness of the nature of the experiment and the exposure undertaken (Wheeler and Alexander, 1992). Even if these problems could be discounted, the problem of differential survival rates between indicator and pathogenic organisms would thwart any extension of the results' validity outside the specific range of hydrographic conditions observed in the experiments.

It has been suggested that the US EPA protocol for epidemiology studies, being unreplicable, failed the first test of an acceptable scientific experiment (Godfree *et al.*, 1990). Failure to achieve any replication even within so small an area as Hong Kong must support this point of view. Furthermore, even if the drastic modifications used in the "healthy volunteer" studies were entirely successful, it could not be concluded that results obtained would be valid elsewhere. The regulator is left to wonder whether the entire edifice of epidemiology might not be dismissed as an unscientific waste of time and effort.

In fact, such a point of view would be unnecessarily pessimistic. The now numerous epidemiology studies do consistently yield certain important, if modest, results:

1. People *report* more illness after swimming (whether it is tested clinically or not does not seem to matter).

2. More specifically, people *report* more illness after swimming if the water is more polluted.
3. *No one indicator* reliably predicts the rate of reporting illness in all circumstances.
4. If the water *looks* dirty, it probably is. High turbidity and the presence of litter may provoke feelings of sickness among swimmers and so increase the rate of reporting symptoms whether or not they have a microbiological cause (Holmes, 1988; Kueh *et al.*, 1995).

These observations, based on a wide variety of epidemiological studies, stress the importance of perceptions and greatly diminish the significance of any one particular indicator of bacterial pollution.

5 PERFORMANCE MEASURES IN POLLUTION CONTROL

The Hong Kong Environmental Protection Department began its epidemiological studies in a rational attempt to assess risks and so to measure and enhance the effectiveness of its performance in water pollution control (Holmes, 1988). In doing so, the importance of community perceptions was recognised, but the objective was to replace misperception with scientific fact.

Now the department has undertaken a very thorough review of its effectiveness in pollution control, intending to devise and adopt new and appropriate measures of performance. Rational, goal-based performance measures are not the only ones that can be considered: there are many alternatives, and no clear consensus on the choice. One such is a systems approach, which uses survival and growth as indicators of success (Holmes, 1994).

Rational goals, amenable to measurement by science, or accountancy, appear to be more appropriate for organizations that are very focused on simple objectives. Because of their narrow, single purpose, such organizations would tend to be vigorous, aggressively pursuing a target, but short-lived, ceasing to have a reason to exist after the target is achieved. Extra-rational goals, such as survival and growth, more realistically measure the performance of organizations that have very broadly defined or uncertain goals. Members of such organizations may still be vigorous, but because their activities are directed in many directions, there is a lesser appearance of strength of purpose in the organization as a whole. Very bureaucratic organizations tend to be self-defensive, growing inward rather than making much difference to the world outside. For such organizations the best measures of performance would relate to work throughput rather than results achieved.

Pollution control organizations would tend to fall into the second category, because of the inherently diffuse nature of goals like "protect the environment" and "protect community health whilst promoting sustainable development". Attempts to devise simple, reliable and universal standards for pollution control in respect of bathing beaches have not yet met with much success: there may be theoretical reasons they can never succeed.

Perceptions, specifically those of the tax-paying public or their political representatives, are the key to determining the effectiveness of pollution control. To return to the regulator's questions, what the public demands to know is not "what is the *degree of pathogenic bacterial contamination* of this bathing water?" nor even "what is the *statistical risk* to my health if I swim here?" It is far more likely to be "is this water *safe* for swimming?" The paying public, in other words, is only interested in outcomes, not in the underlying science. The pollution control authority's effectiveness will be judged by how well it meets the community's irrational expectations.

6 CONCLUSION

Epidemiological studies at bathing beaches in Hong Kong, in common with other studies around the world, have demonstrated that swimmers tend to suffer more illness than those who do not swim. They have also shown that those who swim in more polluted water, as measured by a variety of indicators, report more symptoms of sickness than those who swim in less polluted water. Such studies have not produced a single indicator, bacterial or otherwise, that can predict the rate of occurrence of disease symptoms consistently and reliably, and should not be expected to do so. It is not necessary that they should. Pollution control authorities will be judged by their ability to be seen to improve water quality, in line with community expectations, at the minimum cost, rather than their ability to meet any specific scientific standard.

Acknowledgement

The author thanks the Director of Environmental Protection for permission to publish and present this paper. The opinions expressed are those of the author and do not necessarily represent the official policy of the Hong Kong Government or the Environmental Protection Department.

References

Alkan U, Elliott D J, Evison L M. Survival of enteric bacteria in relation to simulated solar radiation and other environmental factors in marine waters. *Wat. Res.* 1995; **29**(9): 2071-2081.

Anderson T A. Impact in Scotland of UK and EC sewage legislation. *J. Instn. Wat. & Environ. Mangt.* 1992; **6**(6): 682-689.

Cabelli V J, Dufour A P, McCabe L J, Levin M A. Swimming-associated gastroenteritis and water quality. *Am. J. Epidemiol.* 1982; **115**: 606-616.

Cabelli V J. *Health Effects Criteria for Marine Recreational Waters.* U. S. Environmental Protection Agency; 1983. Report number: EPA-600/1-80-031.

Cheung W H S, Chang K C K, *Microbiological Water Quality of Bathing Beaches in Hong Kong 1988.* Hong Kong: Environmental Protection Department; 1988. Report Number: EPD/TR5/88.

Cheung W H S, Hung R P S, Chang K C K, Kleevens J W L. Epidemiological study of beach water pollution and health-related bathing water standards in Hong Kong. *Wat. Sci. Tech.* 1991; **23**: 243-252.

Dupray E, Derrien A. *Influence du passage de* salmonella spp. *et* E. coli *en eaux usées sur leur survie ultérièure en eau de mer. Wat. Res.* 1995; **29**(4): 1005-1011.

Evison L M. Comparative studies on the survival of indicator organisms and pathogens in fresh and sea water. *Wat. Sci. Tech.* 1988; **20**(11/12): 309-315.

Fleisher J M. A reanalysis of data supporting U. S. Federal bacteriological water quality criteria. *Res. J. Water Pollut. Control Fed.* 1991; **63**(3): 259-265.

Fleisher J M, Jones F, Kay D, Stanwell-Smith R, Wyer M, Morano R. Water and non-water-related risk factors for gastroenteritis among bathers exposed to sewage contaminated waters. *Internat. J. Epidem.* 1993; **22**(4): 698-708.

French G L, Ling J, Chow K L, Mark K K. Occurrence of multiple antibiotic resistance and R-plasmids in gram-negative bacteria isolated from faecally contaminated freshwater streams in Hong Kong. *Epidem. Inf.* 1987; **98**: 285-299.

Godfree A, Jones F, Kay D. Recreational water quality - the management of environmental health risks associated with sewage discharges. *Marine Pollution Bulletin* 1990; **21**(9): 414-411.

Hodgkiss I J. Bacteriological monitoring of Hong Kong marine water quality. *Environment International* 1989; **14**: 495-499.

Holmes P R. Standards and risks in the management of water quality in Hong Kong. In: *Pollution In The Urban Environment - Polmet '88*. P Hills *et al.* editors. Hong Kong: the Hong Kong Institute of Engineers, 1988; **2**: 426-431.

Holmes P R. Policies and principles in Hong Kong's water pollution control legislation. *Wat. Sci. Tech.* 1992; **26**(7-8): 1905-1914.

Holmes P R. Bureaucracy and effectiveness in water pollution control. *Wat. Sci. Tech.* 1994; **30**(5): 111-120.

Holmes P R, Smith-Evans M. Monitoring progress towards Hong Kong's water quality objectives. *Aquatic Conservation* 1995; **5**(1):55-65.

Johnstone D W M, Horan N J. Standards, costs and benefits: an international perspective. *J. Instn. Wat. & Environ. Mangt.* 1994; **8**(5): 450-458.

Kay D. Coastal bathing water quality: the application of water quality standards to Welsh beaches. *Applied Geography* 1988; **8**: 117-134.

Kay D, Fleisher J M, Salmon R L, Jones F, Wyer M D, Godfree A F *et al.* Predicting likelihood of gastroenteritis from sea bathing: results from randomised exposure. *Lancet* 1994; **344**: 905-909.

Kueh C S W, Grohmann GS. Recovery of viruses and bacteria in waters of Bondi beach: a pilot study. *Med. J. Aust.* 1989; **151**: 632-638.

Kueh C S W, Tam T Y, Lee T, Wong S L, Lloyd O L, Yu I T S *et al.* Epidemiological study of swimming-associated illnesses relating to bathing beach water quality. *Wat. Sci. Tech.* 1995; **31**(5-6): 1-4.

Lacey R F, Pike E B. Water recreation and risks. *J. Instn. Wat. & Environ. Mangt.* 1989; **3**(1): 13-21.

Mezrioui N, Baleux B, Troussellier M. A microcosm study of the survival of *Escherichia coli* and *Salmonella typhimurium* in brackish water. *Wat. Res.* 1995; **29**(2): 459-465.

Morton B. Pollution of the coastal waters of Hong Kong. *Marine Pollution Bulletin* 1989; **20**(7): 310-318.

Pereira M G, Alcântara F. Culturability of *Escherichia coli* and *Streptococcus faecalis* in batch culture and "in situ" in estuarine water (Portugal). *Wat. Res.* 1993; **27**(8): 1351-1360.

Pike E B. Recreational use of coastal waters: development of health-related standards. *J. Instn. Wat. & Environ. Mangt.* 1993; **7**(2): 162-169.

Tam T Y, Au P T H, Cheung F K, Ko J Y L, Cheung K Y. *Bacteriological water quality of bathing beaches in Hong Kong, 1994*. Hong Kong: Environmental Protection Department; 1994. Report Number: EPD/TR6/94.

Wheeler D, Alexander L M. Assessing the risks of sea bathing. *J. Instn. Wat. Environ.Mangt.* 1992; **6**(4): 459-467.

Wyer M D, Fleisher J M, Gough J, Kay D, Merrett H. An investigation into parametric relationships between enterovirus and faecal indicator organisms in the coastal waters of England and Wales. *Wat. Res.* 1995; **29**(8): 1863-1868.

Treated Waters

Chapter 17
Survival And Growth Of *E. Coli* In Drinking Water Distribution Systems

Block J.C.[1], Mouteaux L.[1], Gatel D.[2,] and Reasoner D.J.[3]

[1] NanCIE/GIP Stelor/LSE, Faculté de Pharmacie, 5 rue Albert Lebrun, F-54000 Nancy.
[2] Compagnie Générale des Eaux , Quartier Valmy, 2 place Ronde, Cedex 82 - F-92982 Paris-La-Défense.
[3] U.S.E.P.A., Office of Research and Development, National Risk Management Research Laboratory, 26 W. Martin Luther King Drive, Cincinnati Ohio 45268, U.S.A.

1 INTRODUCTION

Coliform bacteria and *Escherichia coli* have been used as quality indicators in water distribution systems since the beginning of the century. When detected in drinking water, they indicate fecal contamination and, for the consumer, a risk of infection with pathogenic enteric microorganisms (*Salmonella*, enteric viruses, *Giardia*, etc.). The concept of coliforms as indicators is frequently and regularly the object of discussion since no group of microorganisms fully meets the following theoretical criteria for the perfect indicator (Dutka, 1973) :

1) present in the sample at higher levels than the suspected pathogens,
2) do not grow more or die-off faster than pathogens in the water,
3) more resistant to disinfectants than the pathogens themselves,
4) easily identifiable, and taxonomically unambiguous.

The fact that the coliform bacteria and *E. coli* have been successfully used for many years now would tend to validate the usefulness of this tool. The risk of infections by drinking water with no coliforms or *E. coli* present is very low but rises when they are present (Zmirou *et al.*, 1987). However, since 1930, detection of coliforms in distribution networks (see reviews by LeChevallier, 1990 and Camper, 1994) cannot all be explained by obvious mistakes in the treatment process, broken water mains or accidental contamination. Recurrent episodes of coliforms in some networks or at particular points in the networks and their greater prevalence during warm weather (higher temperatures and higher levels of biodegradable organic matter) (LeChevallier, 1990; Volk and Joret, 1994) suggest that low levels of coliforms have entered the network (Bucklin *et al.*, 1991) and that these bacteria have been growing in the distribution networks.

Direct proof of such growth is still rare and is often based on laboratory evidence (Camper *et al.*, 1993; Mackerness *et al.*, 1991; Packer, 1995; Szewzyk *et al.*, 1994), so much so that the issue of the growth of coliforms, and even more so for *Escherichia coli*, remains debatable.

The purpose of this paper is threefold. First, it will briefly review some points concerning bacterial colonization in distribution networks and the parameters controlling biofilm accumulation. Then it will attempt to demonstrate the ability of *Escherichia coli* to grow in microaerobic media and, lastly it will summarize a study of the behavior of *E. coli* bacteria introduced in a distribution network.

2 BIOFILMS AND WATER DISTRIBUTION NETWORKS

The transit time of treated water between treatment plant output and customer's tap is extremely variable and depends both on the time the readings are made (water consumption is higher and transit time is shorter during the day) and the point in the network (*i.e.* distance from the plant and number of intermediate reservoirs). In practice, transit time varies from 24 hr to one month and this water aging is sufficient to cause physico-chemical and biological reactions which deteriorate the quality of the mains water.

Whatever the characteristics of the treated water at the plant output, all distribution networks are colonized by a live bacterial biomass both in the water and at the surface of the pipes (Block, 1992). The biomass colonizing the surface of the water mains (bacterial, microscopic fungi, yeast cells, protozoa) is in the form of more or less dispersed microcolonies held together by an exopolymer matrix and not a continuous film *per se*. The term biofilm will be used here as a generic term (Figure 17.1).

Bacterial density in the biofilm varies, according to the situation, from 10^3 to 10^7 cells cm^{-2} (*i.e.* 50 to 100 times higher than in the adjacent water phase) of which 1 to 5 % are viable heterotrophic bacteria (able to grow on agar nutrient medium or possessing a functional respiratory chain) even in the presence of disinfectants. This biomass can be described as autochthonous in most cases, *i.e.* perfectly adapted to the water distribution ecosystem, and, as far as we know, very difficult to eradicate.

In many water distribution systems where then are low nutrient concentrations and high water flow rates, bacteria can only grow in the biofilm. Bacterial contamination of the water during transit comes from pieces of biofilm dislodged from the pipe surfaces. This means that any strategy designed to obtain biologically stable water must limit accumulation and proliferation of this fixed biomass.

At least five parameters control biofilm accumulation on water mains: hydraulics of the

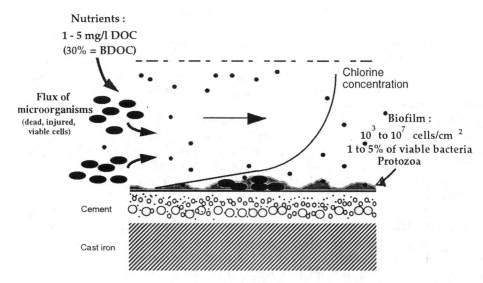

Figure 17.1 *Schematic representation of the biomass colonizing water mains (the thickness of the biofilm is lower than the boundary layer, which is here of 70 µm in thickness).*

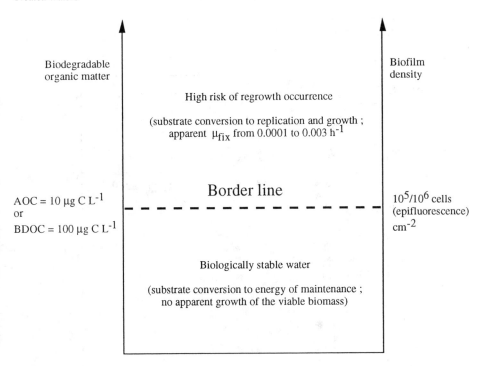

Figure 17.2 *Biodegradable organic matter concentration and substrate conversion.*

system, roughness of the materials, the presence of a disinfectant residual, concentration of nutrients and bacterial density of the water entering the distribution network (see review by Block, 1996). A threshold concentration for available biodegradable organic matter of around 10 µgC L^{-1} could be used if expressed as AOC (Van der Kooij and Veenandaal, 1992; LeChevallier *et al.*, 1987) or 100 µg L^{-1} if expressed as BDOC (Servais *et al.*, 1995), below which the water can be considered as biologically stable (Figure 17.2). However, even such low nutrient concentrations can maintain a permanent viable heterotrophic biomass (Sibille *et al.*, 1995) and can induce accidental coliform proliferation (Block *et al.*, 1995). Lastly, the addition of microorganisms (living/injured, dead) through treated water from the plant (10 to 10^5 mL^{-1}) contributes to continuous network inoculation and biofilm accumulation. Mathieu *et al.* (1993) have suggested that the 2 log increase in bacterial density at the water network input leads to a 1 log increase in the biofilm. Therefore, any accidental addition of coliforms and/or *E. coli* will lead to deposition of these microorganisms in the biofilm depending on the system's hydraulics (Bryers, 1987).

3 *ESCHERICHIA COLI*: A FACULTATIVELY ANAEROBIC BACTERIA

Several authors have isolated coliforms and *Escherichia coli* by scraping sediment and corrosion deposits in water mains (Victoreen, 1984; LeChevallier *et al.*, 1987; Olson *et al.*, 1991). Their presence in such places suggests that *E. coli*, a facultative anaerobic bacteria, can survive and grow even inside corrosion products. This is possible since *E. coli* has very

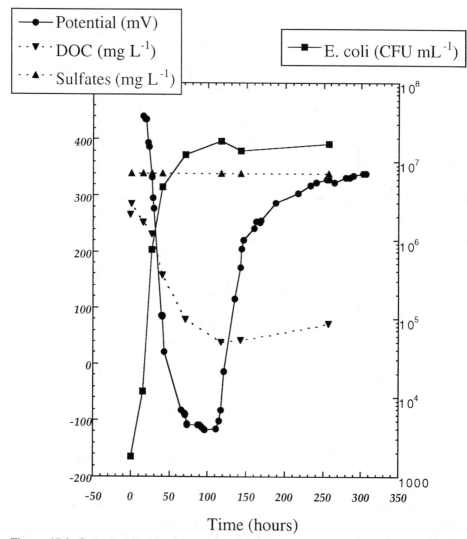

Figure 17.3 *Growth of Escherichia coli and redox potential at 30°C in nutritive medium (Postgate modified). Batch conditions ; no agitation of the 200 mL of medium in the 500 mL erlenmeyer.*

strong affinity for oxygen and can respire nitrates but is unable to use sulfates as electron acceptors (Ingledew and Poole, 1984). In a poorly aerated but nutrient-rich medium (Figure 17.3), we have observed, in agreement with Wimpenny (1969), rapid growth of *Escherichia coli* and oxygen consumption produce a drastic fall in redox potential up to -100 mV without any apparent effect on *E. coli* growth. Furthermore, *E. coli* does not appear to be affected when co-cultured with sulfate-reducing bacteria able to produce H_2S, with a redox potential of -220 mV (Figure 17.4). This clearly shows that different bacterial species can cohabit (sulfate-reducing bacteria + *E. coli*) and gives further evidence of the importance of controlling corrosion to prevent coliform growth (LeChevallier *et al.*, 1993).

4 BEHAVIOR OF *ESCHERICHIA COLI* EXPERIMENTALLY INTRODUCED INTO A DISTRIBUTION NETWORK

The ability of some *Escherichia coli* strains to grow in mains water has been demonstrated in batch tests at temperatures of 20 to 25°C (Byrd *et al.*, 1991; Camper *et al.*, 1991; Rice *et al.*, 1991). Not all strains have the same behavior and, in the same water, initial contamination levels of some strains previously isolated from the environment are maintained while other strains grow significantly (Figure 17.5).

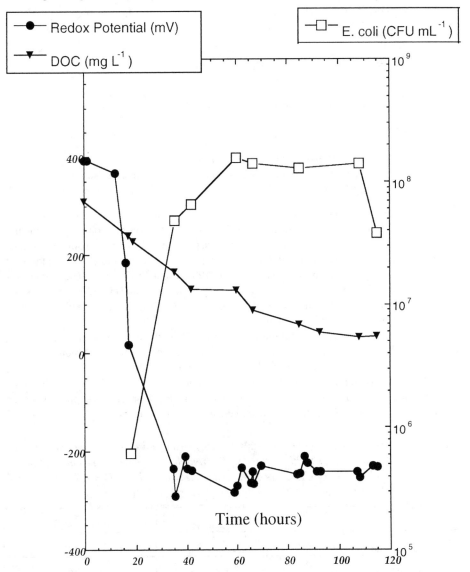

Figure 17.4 *Co-culture of Escherichia coli and sulfate-reducing bacteria at 30°C in a nutritive medium.*

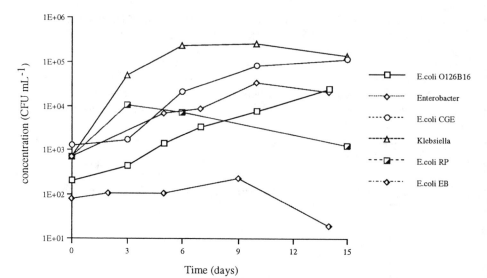

Figure 17.5 *Comparative growth of 5 strains of E. coli and coliforms isolated from*
drinking water distribution systems and 1 strain from a collection (E. coli
O126B16) (Batch test : 300 mL of non-sterile drinking water of Nancy, 20 °C).

It is difficult to experimentally demonstrate colonization of a distribution network by
E. coli, because of the limited sensitivity of the culture detection methods. The following
study was carried out by injecting a dense *E. coli* suspension (10^{11} CFU in a single 5-minute
injection to obtain approximately 10^5 CFU mL^{-1} in the mains water) into one of the loops of
an experimental distribution network (Colin *et al.*, 1987; Block *et al.*, 1993). The *E. coli* was
previously grown in the laboratory in a liquid rich nutrient medium(18h - 37°C), then starved
by incubating for 24 h at room temperature and finally washed in a buffer solution to minimize
the amount of dissolved organic carbon.

The experimental network was continuously supplied with treated plant water for several
weeks (10 L h^{-1}; HRT: 24 hr) and was extensively colonized by a bacterial biomass (Table
17.1). The changes that occur after rapid, single addition of the *Escherichia coli* suspension
can be theoretically summarized by at least three situations described in Figure 17.6 based on
the principle of dilution in a perfectly mixed reactor fed with a continuous supply. However,
the experimental results (Figure 17.7) describe a more complex response:

Table 17.1 *Microbiological characteristics of the experimental network continuously fed*
for several weeks with finished water from the Nancy water treatment plant (n
= 7) (average temperature 20 °C ; chlorine = 0 mg L^{-1} ; pH = 8.7 ; DOC = 1.7
mg L^{-1} ; hydraulic residence time = 24 hours)

	Biofilm (cm^{-2})	Water (mL^{-1})
Total cells (epifluorescence)	4.7 10^6	5.2 10^5
CFU-15 days	2.1 10^5	8.1 10^3
CFU-3 days	0.6 10^5	0.4 10^3

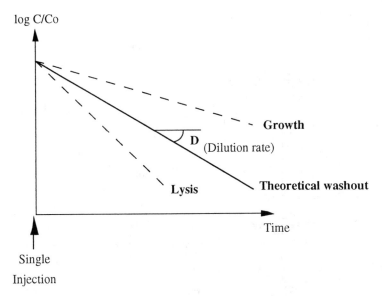

Figure 17.6 *Theoretical behaviour of Escherichia coli suspension introduced in a perfectly mixed reactor continuously fed with drinking water*

- First, within a few hours of introducing bacterial pollution, a fraction of the contaminating organisms will adhere to the pipe surface. From results of studies not reported in this paper, 1 to 10 % of the added bacteria fixed onto the surfaces within 3 hours. During this initial phase, the density of *Escherichia coli* was higher in the water phase than in the biofilm.

- Second, 7 to 8 days after contamination, the slope of the *E. coli* biofilm concentration curve changed and cut clearly across *E. coli* disappearence curve for the water ; by day 12, the biofilm curve crossed the theoretical washing curve. This phenomenon can only be due to *Escherichia coli* growth in the biofilm. However, we were unable to obtain in any of our tests a plateau for or stabilization of *Escherichia coli* counts. Growth of these bacteria was slightly lower than the dilution rate in the system and their best growth rate (μ), calculated according to the method of Van der Wende *et al.* (1989), was around 0.002 h^{-1}, *i.e.* an apparent doubling time of approximately 14 days at 20°C. Consequently, colonization of the network by *E. coli* was only partial and transient despite their ability to grow. In different studies not reported in this paper, despite their growth, *Escherichia coli* were no longer detectable - less than 1/100 mL - 5 to 6 weeks after contamination.

- Third, the curves in Figure 17.7 show an inversion in the total amounts of *E. coli* in the water and biofilm. After 10 days, the biofilm system contained approximately 10 times more *E. coli* than the water phase. This clearly shows preferential *Escherichia coli* growth in the biofilm.

5 CONCLUSIONS

The *Escherichia coli* strain used in this experimental study grew at 30°C in co-culture with sulfate-reducing bacteria (*i.e.* in a slightly anaerobic medium) and/or in a biofilm of viable heterotrophic aerobic bacteria in a drinking water distribution network. There did not appear

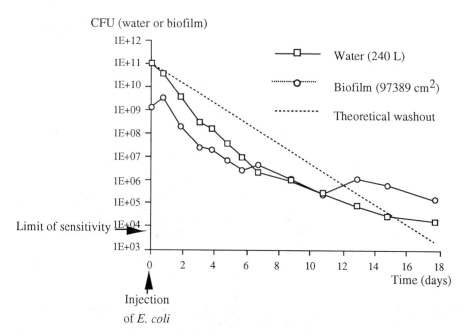

Figure 17.7 *Behavior of Escherichia coli experimentally introduced in the water distribution system. (One single injection of 10^{11} CFU ; D = 0.042 h^{-1} ; temperature : 20°C. The data are expressed as the total number of CFU present at time t in the water of the system, 240 liters ; or at the surface of pipes : 97,389 cm^2 of cement lined cast iron pipe).*

to be any drastic competition for nutrients between the indigenous bacterial biomass already in place and the contaminating *E. coli*. In most distribution networks, the biodegradable organic carbon has been found in measurable amounts throughout the network (Mathieu *et al.*, 1995).

The growth rate of *E. coli* depends not only on the strain but most of all on the availability of biodegradable organic matter. The *E. coli* used to contaminate this experimental distribution network had variable growth rates (*i.e.* 14-day doubling times gradually getting longer) and consequently were unable to permanently colonize the network. This unstable state or transient colonization undoubtedly explains why *Escherichia coli* could only be detected in an erratic manner in the distribution networks. It should be kept in mind that, even after massive contamination, in 6 to 7 days at 20°C, only a small fraction of the *Escherichia coli* suspension injected (*i.e.* 0.001 %) was able to adapt and start to grow. It is far from certain that even transient colonization would occur with lower contamination in under winter cold water conditions and low nutrient concentration. The model of Volk and Joret (1994) shows that the best chance of isolating coliforms in distribution networks is during the "warm season" *i.e.* when water temperature exceeds 15°C , therefore stimulating biodegradation of organic substances by the network biomass (consumption of more than 0.15 mg BDOC L^{-1}) and causing intensive growth (bacterial cells > $10^{5.2}$ mL^{-1}).

It would be unreasonable today to reject *Escherichia coli* as an indicator of fecal contamination because it can, in some circumstances, proliferate in the biofilms of drinking water distribution networks. This potential growth capacity does not call into question the type or source of fecal contamination but makes it impossible to diagnose or judge when contamination started. This contamination can sometimes be quite old (*i.e.* date back several weeks or even several months).

Apart from the facts confirming that enteric bacteria can grow in a rather dilute and cold environment (compared to the digestive tract of mammals), we should address the issue of the behavior of other coliforms and opportunistic bacteria placed under the same conditions. These facts could perhaps then explain the links between gastrointestinal diseases and drinking water consumption (Payment *et al.*, 1991 a and b).

Acknowledgements

The results presented in this paper are part of the research programme driven by of the Biofilm Research Group: NANC.*I.E.*; C.G.E.; S.E.D.I.F.; Agence de l'Eau Seine Normandie; U.S.E.P.A.

References

Block J.C. Biofilms in drinking water distribution systems. In Melo L., Bott T.R., Fletcher M. Capdeville B. editors. NATO Advances Study Institute, Portugal, Kluver, 1992: 469-485.

Block J.C. Biodegradable organic matter in water distribution systems. In Prévost M., Joret J.C. editors, "Biodegradable organic matter in drinking water", Lewis, New York, 1996.

Block J.C., Gatel D., Reasoner D.J., Lykins B., Clark R.M. Biodiversity in drinking water distribution systems : a review. In Microbiological quality of water, IWSA/FBA Conference, London, 12-13 December 1995.

Block J.C., Haudidier K., Paquin J.L., Miazga J. and Lévi J. Biofilm accumulation in drinking water distribution systems. Biofouling, 1993;6: 333-343.

Bryers J.D. Biologically active surfaces : processes governing the formation and persistence of biofilms. Biotechnol. Progress 1987;3: 57-68.

Bucklin K.E., McFeters G.A. and Amirtharajah A. Penetration of coliforms through municipal drinking water filters. Wat. Res. 1991;8: 1013-1017.

Byrd J.J., Xu H.S. and Colwell R.R. Viable but nonculturable bacteria in drinking water. Appl. Environ. Microbiol. 1991;57: 875-878.

Camper A.K. Coliform regrowth and biofilm accumulation in drinking water systems : a review. In Geesey G.G., Levandowski M. and Flemming H.C. (eds), Biofouling and biocorrosion in industrial water systems, Lewis Publishers, N.Y., USA, 1994:91-105

Camper A.K., McFeters G.A., Characklis W.G. and Jones W.L. Growth kinetics of coliform bacteria under conditions relevant to drinking water distribution systems. Appl. Environ. Microbiol. 1991;57: 2233-2339.

Camper A.K., Hayes J.T., Jones W.L. and Zelver N. Persistence of coliforms in mixed population biofilms. Proceedings of the Water Quality Technology Conference; 1993 November 7-11; Miami, American Water Works Association, Denver.

Colin F., Grapin G., Chéron J., Lévi Y., Pozzoli B., Miazga J. *et al.*. Etude de l'évolution de la qualité de l'eau potable dans les réseaux de distribution : une approche et des moyens nouveaux. Techn. Sci. Munic. Eau. 1987;12:565-574.

Dutka B.J. Coliforms are an inadequate index of water quality. J. Environ. Hlth. 1973;36: 29-46.

Ingledew W.J. and Poole R.K. The respiratory chains of *Escherichia coli*. Microbiol. reviews, 1984;48: 222-271.

LeChevallier M.W. Coliform regrowth in drinking water : a review. J. Am. Wat. Wks Ass. 1990;82:74-86.

LeChevallier M.W., Lawry C.D., Lee R.G. and Gibbon D.L. Examining the relationship between iron corrosion and the disinfection of biofilm bacteria. J. Am. Wat. Wks Ass. 1993;85:111-123.

LeChevallier M.W., Babcock T.M. and Lee R.G. Examination and characterization of distribution system biofilms. Appl. Environ. Microbiol. 1987;53:2714-2724.

Mackerness C.W., Colbourne J.S. and Keevil C.W. Growth of *Aeromonas hydrophila* and *Escherichia coli* in a distribution system biofilm model. Proceedings of the UK Symposium on health related water microbiology, Glasgow, UK; 1991:131-138

Mathieu L., Block J.C., Dutang M., Mailliard J. and Reasoner D.J. Control of biofilm accumulation in drinking-water distribution systems. Water Supply. 1993;11:365-376.

Mathieu L., Block J.C., Prévost M., Maul A. and De Bishop R. Biological stability of drinking water in the city of Metz distribution system. Aqua J. Water-SRT. 1995;44:230-239.

Olson B.H., McCleary R. and Meeker J. Background and models for bacterial biofilm formation and function in water distribution systems. In C.J. Hurst (ed.), Modelling the environmental fate of microorganisms, A.S.M. Publishers, Washington DC, USA. 1991:255-285.

Packer P.J. Does *Klebsiella oxytoca* grow in the biofilm of water distribution systems ? Int. Conference on "Coliforms and *E. coli* : problem or solution", University of Leeds (UK) 1995.

Payment P., France E., Richardson L. and Siemiatycki J. Gastrointestinal health effects associated with the consumption of drinking water produced by point-of-use domestic reverse-osmosis filtration units. Appl. Env. Microbiol. 1991a;57:945-948.

Payment P., Richardson L., Siemiatycki J., Dewar R., Edwardes M. and France E. A randomized trial to evaluate the risk of gastrointestinal disease due to the consumption of drinking water meeting currently accepted microbiological standards. Amer. J. Public Health. 1991b;81:703-708.

Rice E.W., Scarpino P.V., Reasoner D.J., Logsdon G.S. and Wild D.K. Correlation of coliform growth response with other water quality parameters. J. Am. Wat. Wks. Ass. 1991;83:98-102.

Servais P., Laurent P. and Random G. Comparison of the bacterial dynamics in various French distribution systems. J. Water SRT-Aqua. 1995;44:10-17.

Sibille I., Mathieu L., Paquin J.L., Hartemann Ph., Clark R., Lahoussine V. *et al.* Improvement of water quality during distribution using nanofiltration process. AWWA *"Water quality technology conference"*, 1995 November, Louisianna, USA.

Szewzyk U., Manz W., Amann R., Scheilfer K.H. and Strenström T.H. Growth and *in situ* detection of a pathogenic *Escherichia coli* in biofilms of a heterotrophic water-bacterium by use of 16S- and 23S-r RNA-directed fluorescent oligonucleotide probes. FEMS Microbiol. Ecol. 1994;13:169-176.

Van der Kooij D. and Veenendaal H.R. Assessment of the biofilm formation characteristics of drinking water. AWWA WQTC, Toronto, Cd 1992.

Van der Wende E., Characklis W.G. and Smith D.B. Biofilms and bacterial drinking water quality. Wat. Res. 1989;23:1313-132.

Victoreen H.T. The role of rust in coliform regrowth, AWWA WQTC Denver,Colo. 1984:253-264.

Volk C. and Joret J.C. Paramètres prédictifs de l'apparition des coliformes dans les réseaux de distribution d'eau d'alimentation. Sciences de l'Eau. 1994;7:131-52.

Wimpenny J.W.T. The effect of Eh on regulatory processes in facultative anaerobes. Biotechnol. Bioengng. 1969;11:623-629.

Zmirou D., Ferley J.P., Collin J.F., Charrel M. and Berlin J. A follow-up study of gastro-intestinal diseases related to bacteriologically substandard drinking water. Am. J. Public Health. 1987;77:582-584.

Chapter 18
Predicting Coliform Detection In The Drinking Water Supply

P. Gale and R.F. Lacey

WRc plc, Henley Road, Marlow, Buckinghamshire, SL7 2HD, UK.

1 INTRODUCTION

Bacteria belonging to the group called total coliforms are indicators of microbiological water quality. Under The Water Supply (Water Quality) Regulations (1989) for England and Wales and The Water Supply (Water Quality) (Scotland) Regulations (1990) for Scotland, coliforms should be absent in more than 95% of 100 ml samples taken from consumer premises in each water supply zone. The UK water industry has devoted considerable research and operational effort to improving the compliance of zones with this target. That commitment has been rewarded in a steady increase in compliance each year from 1990 when only 91% of supply zones in England and Wales complied with the coliform standard. Zonal coliform compliance increased to 94.8% in 1991, 96.7% in 1992 and almost 99% in 1993. In 1994, 99.5% of supply zones complied (Department of Environment, 1995). However, intermittent occurrences of coliforms are still of concern to operators of water supply undertakings. This work was funded by the UK water industry through the management of UK Water Industry Research (UKWIR) Ltd..

The general objective of this work is to shed light on the reasons why coliforms occur. More specifically, the objectives are:-

- To identify statistical associations, both at the sample and zonal level, between coliform occurrences and other routine determinands in statutory monitoring data provided by water supply undertakings.
- To develop and validate simple algorithms for predicting increased probability of coliform failures.

1.1 Potential benefits

Identifying associations between the presence of coliforms and fluctuations in other factors at the zone and sample level will facilitate identification of root causes and origins of coliforms so that operational strategies may be developed to minimise their occurrence. Developing models to predict the probability of coliform detection on statutory 100 ml volume sampling would benefit the water industry since remedial treatments could be effectively targeted to specific regions of the supply before zones fail to comply.

1.2 General Approach

Water supply undertakings in the UK have established large water quality databases through statutory monitoring under the water quality regulations. Those databases contain information for coliforms and chemical, physical and other microbiological factors for the same water samples. Statistical analyses performed previously on those databases showed that water samples with coliforms had higher mean temperatures and plate counts but lower mean total chlorine concentrations than samples without coliforms (Gale, 1994a).

The statistical method used in the present work is based on the concept of a 'generalised linear model' or GLM (McCullagh and Nelder, 1989). This choice of method was made because of the ability of GLM to meet the following requirements:-

- to enable the probability of coliform occurrence to be made the subject of prediction;
- to enable the effects of several explanatory variables (e.g. total chlorine, temperature and plate count) to be modelled simultaneously.
- to distinguish associations of coliforms with other factors which operate *within-zone* (i.e., locally in space and time between different 100 ml volume samples) from those which operate *between-zone* (i.e., due to systematic differences from zone to zone).

Within-zone associations are reflected in higher or lower total coliform densities at certain regions of the zone or at certain times of the year. Between-zone associations are reflected in higher or lower total coliform densities in different zones. Factors related to the quality of the source water and its treatment would be candidates for between-zone associations. Some factors, e.g. the degree of chlorination, may exhibit effects on coliforms at both levels (i.e. within-zone and between-zone).

2 STATISTICAL METHODS

Statutory water quality monitoring data collected during 1992 or 1993 were provided by ten water supply undertakings in the UK. Data for confirmed coliforms (Department of Environment *et al.*, 1983) were used. Within each water undertaking, zones were categorised according to treatment works, disinfection type, or source water prior to analysis by GLM.

2.1 Generalised Linear Modelling

GLM was performed using the Genstat statistical package (Genstat 5 Committee, 1993). A version of GLM based on the Poisson statistical distribution to predict coliform densities resulted in many data points with large standard residuals and was therefore judged as inappropriate. This may reflect the heterogeneous distribution of bacteria (Maul *et al.*, 1985) including coliforms (Christian and Pipes, 1983; Gale, 1994b) in drinking water supplies. A version of the GLM approach which expresses the probability of coliform detection directly as a function of the external factors so that

p (coliform present) = f (temperature, total chlorine, plate count)

was used. The equation is written as:-

$$\ln (p/(1 - p)) = a_0 + a_1.\text{Temperature} + a_2.[\text{Total Chlorine}]$$
$$+ a_3.[\text{Plate count}] \qquad \text{Equation 1}$$

where p is the probability of coliform detection in a 100 ml volume and a_0, a_1, a_2 and a_3.... are constants whose values are estimated from the data. To allow for possible differences between zones the general constant, a_0, can be replaced by a zone-specific constant, a_{zone}. This enables 'zone' to be treated as a qualitative 'factor' in the GLM, in a very similar way to the continuous variables; temperature, total chlorine and plate count.

This model assumes a binomial statistical distribution for coliform occurrences and fitting therefore requires data with coliform-negative samples as well as coliform-positive samples. This is appropriate for 100 ml volume drinking water samples where ~95% of samples register 0 coliforms. The fitting procedure assumes that the detection of coliforms is always based on examination of the same sample volume (100 ml).

Zonal differences in coliform occurrences were tested with just 'zone' as a factor in the model. Temperature, total chlorine and plate count were added to the model. If the GLM analysis shows that a coefficient a_1, a_2, or a_3 is statistically (significance level of 5%) different from zero then it is concluded that the presence of coliforms is associated with fluctuations in the corresponding factor. Within-zone and between-zone associations were tested by comparing the effects on the coefficients a_1, a_2, and a_3 of dropping zone as a factor from the model. If dropping 'zone' as a factor from the model had no statistically significant diminution on the predictive power, the model was judged to be complete in that zonal coliform differences were accounted for by total chlorine and/or plate count. For incomplete models a constant, a_{zone}, which is specific for each zone, must be retained, in place of the general constant a_0 in Equation 1.

2.2 Validation of models for predicting the probability of detecting coliforms

Validation is an important step in establishing the applicability of models to systems other than those from which the models were derived, or their applicability to the same systems in different periods of time. GLM analysis was performed on 1993 data from each water supply undertaking to produce a model like Equation 1. Transforming Equation 1, the probability, p, that a sample contains one or more coliforms is given by:-

$$p = e^R/(1 + e^R) \qquad \qquad \textbf{Equation 2}$$

where $R = a_0 + a_1.\text{Temperature} + a_2.[\text{Total Chlorine}] + a_3.[\text{Plate count}]$ and is the right hand side of Equation 1. Using the values of temperature, total chlorine and plate count for each sample in the corresponding 1991 and 1992 data sets, Equation 2 was used to predict the probabilities of coliform occurrence. In general the predicted probabilities are very low (less than 0.1) for each sample. Because of this, validation cannot be performed at the individual sample level. Validation was therefore assessed in terms of ability to predict the overall number of coliform positives within samples from a group of related zones over the period of one year.

For such a set of samples, the chi-squared test was used to compare statistically the observed number, C, of coliform positive samples with the so-called 'expected' number, E, predicted by the model. This expected number is defined as the sum of the predicted p values when the model is applied to the observed values of the explanatory variables (temperature, total chlorine and plate count) for each sample in turn.

3 RESULTS

3.1 Associations of coliform occurrence with other factors

The presence of coliforms was found by GLM analysis to be consistently associated with higher temperatures, higher numbers of plate count and lower concentrations of chlorine.

3.1.1 Plate count In all of the eight water supply undertakings with three day (22°C) plate count data, the probability of coliform detection increased with increasing numbers of three day plate count. Coliform occurrences were also associated with higher one day (37°C) plate count in all of the six undertakings providing data. Associations of coliforms with two day (37°C) plate count were not universal being statistically significant in three out of five undertakings. All associations between coliforms and plate count operated at the sample level, i.e. within-zone. Thus, plate count appears to be a good predictor of local deterioration in water quality (with respect to coliforms) within a zone, but not a good predictor of which zones within a water supply undertaking are of poorer quality with respect to coliforms.

3.1.2 Total Chlorine concentration Here, coliforms have been modelled with total chlorine irrespective of whether chloramine or free chlorine was used as disinfectant. In eight out of the ten water supply undertakings, the probability of coliform detection increased with decreasing total chlorine concentration. For two undertakings, the effect of total chlorine operated exclusively between-zone, i.e., coliforms were more likely across zones with lower zonal total chlorine concentrations. For four undertakings the effect of total chorine operated exclusively at the sample level, i.e., within-zone.

3.1.3 Temperature In three out of four undertakings which provided temperature data, the probability of coliform detection increased with increasing temperature. The effect of temperature operated within-zone at the sample level, i.e. samples collected within-zone during warmer seasons were more likely to contain coliforms.

3.2 Completeness of Coliform Predictive Models

Differences in coliform probability between-zone were observed within six of the ten water undertakings studied. In five of those undertakings zonal coliform differences could not be fully explained by total chlorine, temperature and/or plate count. Those models are therefore incomplete in the sense that further zonal information (contained in the constant a_{zone} which is retained) is required. In the case of a small undertaking supplying seven zones, differences in total chlorine concentration were alone sufficient to fully explain the observed zonal differences in coliform occurrences.

3.3 Validation of Models: Predicting zonal compliance for previous years

Models developed using 1993 statutory monitoring data were validated on the corresponding 1991 and 1992 data for two undertakings (I and II). Those undertakings were chosen because they comprised a relatively small number of zones with similar properties. The coefficients of the models for undertakings I and II based on 1993 data are shown in Table 18.1. It is apparent that there is not a great difference between these models. Validation exercises were therefore carried out to investigate the following questions:

1. How well can the separate undertaking models for 1993 predict the numbers of coliform positive samples in 1991 and 1992?
2. Could the same model successfully predict the numbers of coliform positive samples in both undertakings?

Parameter	Undertaking I	Undertaking II	Average
Constant	-3.1 (0.5)	-2.4 (1.0)	-2.75
Total Chlorine	-15.0 (5.7)	-11.2 (5.2)	-13.1
Two day PC	0.012 (0.002)	0.0096 (0.008)	0.011

Table 18.1 *Estimates of coefficients (with standard errors) for models of Undertakings I and II based on GLM analysis of 1993 data*

The observed number of coliform occurrences in 1991 and 1992 are compared with those predicted by the 1993 model for water undertakings I and II in Table 18.2 and Table 18.3, respectively. With the exception of the over-prediction for 1991 in undertaking I, the observed and predicted numbers of coliform-positive samples show remarkably good agreement, and for the type of model could not be expected to be closer. However, for both undertakings, the predictions based on the average model fit less well to the observed values (Tables 18.2 and 18.3). Validation of a single model for both undertakings has therefore failed. This may not be too surprising in the case of those two undertakings because:-

•   Undertaking I used free chlorine as disinfectant while undertaking II used chloramine. The total chlorine terms in the respective models therefore reflect different molecular species.
•   The effect of total chlorine on coliforms operated within-zone in undertaking I but between-zone in undertaking II.

In Table 4, the observed and predicted numbers of coliform occurrences for undertaking II are broken down by zone for 1991 and 1992. There is good agreement and in particular, the higher number of coliform occurrences in Zone 2 in 1991 is predicted by the 1993 model.

3.4 Applications of coliform predictive models to the water industry

3.4.1 Prediction of probability of coliform occurrences (per 100 ml) in a region of the supply. Most samples taken from the drinking water supply register 0 coliforms per 100 ml. The true coliform densities in those regions, however, are unknown but could be as low as 1 per 1,000,000 litres or as high as 1 per litre. Operators do not know how close each 0 per 100

	1991	1992	1993
Observed	12	22	24
Predicted by own model	24.7	20.7	24.3
Predicted by average model	35.8	30.3	34.6

Table 18.2 *Total numbers of samples with coliforms, observed and predicted by different models in undertaking I*

	1991	1992	1993
Observed	18	11	7
Predicted by own model	15.3	12.7	7.0
Predicted by average model	9.7	7.8	3.7

Table 18.3 *Total numbers of samples with coliforms, observed and predicted by different models in undertaking II*

Zone	1991		1992		1993	
	Obs	Ex	Obs	Ex	Obs	Ex
1	2	3.1	2	2.5	3	1.8
2	7	5.5	2	1.7	2	1.2
3	1	2.5	1	1.5	2	1.2
4	1	1.4	1	1.5	0	0.7
5	4	1.2	2	1.5	0	0.6
6	2	0.7	0	2.3	0	0.7
7	1	0.9	3	1.7	0	0.8
TOTAL	18	15.3	11	12.7	7	7.0

Table 18.4 *Observed and predicted numbers of coliform-positive samples in undertaking II in 1991, 1992 and 1993 (Model calibrated on 1993)*

ml sample was to registering a coliform. Large volume sampling would provide further information. As an alternative approach coliform predictive models may be used to predict the probability of coliform detection in a 100 ml volume on the basis of plate count, total chlorine, and temperature measured for the sample. An example of a complete model for 89 zones in an undertaking is:-

$$\ln (p / (1 - p)) = -5.3 + 0.0879.\text{Temperature } (°C) + 0.0078[3 \text{ day plate count (no/ml)}] - 2.83[\text{Total Chlorine (mg/l)}] \qquad \textbf{Equation 3}$$

From Equation 2, the probability, p, of detecting a coliform in a 100 ml volume is calculated for extreme winter and summer scenarios (Table 18.5). The model predicts that in the winter scenario coliform detection in 100 ml volumes is unlikely, while in summer scenario the zone would fail to comply to the water quality regulations with a predicted 7.3% coliform-detection rate. Using the annual averages for the 89 zones in 1993, the model predicts that 0.7% of 100 ml volumes would be coliform-positive (Table 18.5).

*3.4.2 Estimating the chlorine concentrations at which probability of coliform detection is less than 5%.*Using the coliform predictive model in Equation 3, a plot such as that shown in Figure 1 may be constructed which relates the total chlorine, plate count and temperature values for which the probability of coliform detection is 5%; above which a zone will fail to comply with the water quality regulations. Total chlorine concentration, unlike plate count and temperature, may be controlled operationally. From Figure 18.1, the total chlorine concentration needed to maintain coliform detection below 5% at different temperatures and plate count density may be determined. The model shows that in winter conditions (e.g. 5°C), chlorination is not necessary for zonal coliform compliance until plate counts exceed 250 per ml. In summer when water temperatures may reach 25°C, at least 0.6 mg/l total chlorine is required when plate count numbers exceed 250 per ml

	Winter	Average	Summer
Temperature (°C)	5	13.31	20
3 day plate count (no/ml)	0	16.9	200
Total Chlorine (mg/l)	0.9	0.3202	0.2
Predicted Probability	0.0006	0.0073	0.0726

Table 18.5 *Predicted probabilities of coliform detection (per 100 ml) in 89 related zones under summer and winter extreme scenarios.*

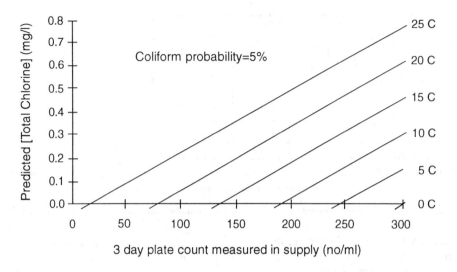

Figure 18.1 *Total chlorine concentrations predicted to maintain confirmed coliform detection rate at 5% from measured three day plate count and temperature.*

4 CONCLUSIONS

1. Using statutory water quality monitoring data, GLM has proved a successful method for modelling the probability of detecting coliforms in 100 ml volume samples taken from specific groups of zones within water supply undertakings.
2. Coliform predictive models have been developed based on temperature, total chlorine concentration and/or plate count.
3. Some models are incomplete, in the sense that information, in addition to that provided by plate count numbers and total chlorine data, is required to explain observed differences in coliform detection rates between zones. Those differences are yet to be characterised, and may include information additional to that available from statutory monitoring.
4. The models developed here are already capable of cautious application within the zones in which they were calibrated, but not yet of extrapolation to other systems.

Acknowledgements

We thank the UK water industry for funding this work through the management of UK Water Industry Research (UKWIR) Ltd. and the water undertakings for providing data. This article is published by kind permission of UKWIR and all copyrights remain the property of UKWIR.

References

Christian RR, Pipes WO. Frequency distribution of coliforms in water distribution systems. Applied and Environmental Microbiology 1983; 45:603-609.
Department of the Environment, Department of Health and Social Security and Public Health Laboratory Service. The bacteriological examination of drinking water supplies 1982.

Reports on public health and medical subjects No. 71. London: HMSO, 1983.

Department of Environment. Drinking water 1994. A report by the chief inspector Drinking Water Inspectorate. London: HMSO, 1995.

Gale P. Statistical associations of external parameters with coliform failures in distribution systems using statutory monitoring data. Marlow (UK) Foundation for Water Research; 1994a. Report No.:FR 0436.

Gale P. Comparison of the proportion of coliform-positive drinking water samples with statistical parameters for log-normal distributions describing coliform concentrations within the supply. Marlow (UK) Foundation for Water Research; 1994b. Report No.:FR 0462.

Genstat 5 Committee. Genstat 5 release 3 reference manual. Oxford: Oxford University Press, 1993.

Maul A, El-Shaarawi AH, Block JC. Heterotrophic bacteria in water distribution. I. Spatial and temporal variation. The Science of the Total Environment 1985; 44:201-214.

McCullagh P, Nelder JA. Generalised linear models. London: Chapman & Hall, 1989.

The water supply (water quality) regulations. Statutory instruments, No. 1147, Water, England and Wales. London: HMSO, 1989.

The water supply (water quality)(Scotland) regulations. Statutory instruments, No. 119 (S.11), Water Supply, Scotland. Edinburgh: HMSO, 1990.

Chapter 19
The Public Health Significance Of Bacterial Indicators In Drinking Water

Martin J. Allen, AWWA Research Foundation*

Stephen C. Edberg, Yale University

1 INTRODUCTION

In the United States, Canada, and many other countries, bacteriological indicators are employed to assess if drinking water is free of infectious organisms. The World Health Organization's *Guidelines for Drinking-water Quality* (World Health Organization, 1993) states that it is impractical to monitor drinking water for every possible microbial pathogen, and that a more logical approach is the detection of organisms normally present in the feces of humans and other warm-blooded animals.

Since the late 1800s, public health scientists have looked for the optimal indicator for drinking water which could be used routinely to assess if drinking water is safe to drink. The indicator should provide a true assessment of the probability of pathogens in drinking water, be relatively inexpensive and easy to assay, and provide definitive results in the shortest period of time.

Three indicator groups are currently used for monitoring drinking water. Unfortunately, the lack of a fundamental understanding on the public health significance of the indicator groups among users and public health agencies has resulted in regulations based on the less specific indicators, which frequently results in inappropriate public health advisories to consumers.

2 INDICATOR GROUPS

2.1 Coliform Group

As early as 1905, public health scientists recommended (American Public Health Association, 1905) the organism *Bacillus coli, i.e., Escherichia* coli, as the preferred indicator of fecal contamination since this organism was universally present at high numbers in both human and animal feces. Unfortunately there was no method that specifically enumerated this species at that time. For this reason, the coliform group, also called "total coliforms," was developed as the surrogate for *E. coli*.

The coliform group comprises genera that satisfy a functional definition, *i.e.*, they utilize lactose to produce acid and gas, or possess the enzyme β-D-galactosidase, which is capable of using a chromogenic galactopyranoside substrate for growth. The genera that satisfy the definition are:

- *Klebsiella* - may be found in feces and ubiquitous in the environment
- *Escherichia* - always found in human and other animal feces
- *Enterobacter* - may be found in feces and ubiquitous in the environment
- *Citrobacter* - found in environmental sources
- *Serratia* - found in environmental sources

It is important to understand that four of these genera are widely found in the environment (source waters, vegetation, soils), are not associated with fecal contamination, and do not pose nor necessarily indicate a health risk. Studies (AWWA Research Foundation, 1987; AWWA Research Foundation, 1989; Edberg *et al.*, 1994) have documented that *Enterobacter* and *Klebsiella* frequently colonize the interior surfaces of water mains and storage tanks (often called "regrowth"), growing in biofilms when conditions are favorable, i.e., nutrients, warm temperatures, low disinfectant concentrations, long residence times, etc.

As the biofilm grows, there is intermittent shedding that can result in more extensive colonization of the distribution system. Obviously, if this situation occurs, a large number of water samples could be coliform positive, triggering non-compliance with certain regulations, and in many instances, needless public advisories for consumers to boil their drinking water. This situation is not uncommon in North American with several hundred water utilities annually encountering regrowth problems. In these situations, the usefulness of the coliform test is lost until the regrowth/colonization situation is controlled. Unfortunately, existing regulations compel water utilities to advise tens of thousands of consumers that their drinking water may be unsafe, even though the causative bacteria neither pose nor indicate a health risk.

2.2 Fecal (Thermotolerant) Coliforms

Since coliforms other than *E.*coli are widespread in the environment, the fecal coliform test or "elevated temperature" test was developed in 1904 to screen for *E. coli*. The selection of the term "fecal" as the name of this method was in retrospect, a very poor choice, since this implied that all fecal coliforms originated from feces; this is incorrect.

In theory, the 44.5°C incubation temperature inhibits the growth of non-thermotolerant coliforms other than *E. coli*. *Citrobacter* and *Enterobacter* are most often eliminated by the elevated temperature, but a significant percentage of *Klebsiella* are thermotolerant. One study (Capelnas and Kanarek, 1984) reported that 15 percent of *K. pneumonia* are thermotolerant, and others (Edberg *et al.*, 1994) reported similar findings shown below.

Table 19.1 *Fecal Coliform Thermotolerance Methods: Species Isolated from Water.*
Source: AWWA Research Foundation Comparison of the Colilert Method and Standard Fecal Coliform Methods, 1994. 90647

| Species | Percent of all species identified | | |
	EC broth	m-FC plate Light Blue	Dark Blue
Escherichia coli	81	64	79
Klebsiella pneumoniae	9	13	10
Klebsiella oxytoca	8	18	9
Enterobacter cloacae	1	3	1
Citrobacter diversus	1	2	1

The fecal coliform test must be performed under exacting temperature standards. Incubation temperatures must be rigidly controlled since even minor excursions as small as 0.2°C will produce erroneous data. Also, large populations of heterotrophic bacteria interfere with both liquid (MPN) and membrane methods. More alarming is the fact that many strains of *E. coli* are unable to ferment lactose or are not thermotolerant (Dufour, 1977), resulting in a false-negative reaction.

Since the current U.S. and Canadian regulations/guidelines allow the use of the fecal coliform method, many water utilities are using a method that enumerates non-fecal *Klebsiella* and actually inhibits the growth of fecal-associated *E. coli*. For these reasons, the fecal coliform method for compliance should be reevaluated.

While a significant percentage of *Klebsiella* isolates grow at 44.5°C (fecal coliform positive), this genus does not pose a health risk (Duncan, 1988). In an extensive review entitled "Waterborne *Klebsiella* and Human Diseases," I.B.R. Duncan concluded that "....*Klebsiella* in water supplies should therefore not be considered a hazard to human health." Other studies have isolated *Klebsiella, Enteroboacter,* and *Citrobacter* in large numbers from sapwood of a variety of trees, concluding that these organisms are indigenous to the wood (Bagley *et al.*, 1978).

This lack of specificity for accurately differentiating between fecal and nonfecal sources of these genera compromises the value of the fecal coliform method for assessing drinking water quality.

2.3 *Escherichia coli*

Starting in the late 1980s, a number of studies (Edberg *et al.*, 1988; Edberg *et al.*, 1989, Rice *et al.*, 1990; Covert *et al.*, 1992) reported the successful use of defined substrate media to selectively enumerate *E. coli*. This medium contained 4-methyl-umbelliferyl-β-D-glucuronide (MUG) that can only be metabolized by the unique constitutive enzyme, β-glucuronidase, found in at least 95 percent of *E. coli*. After extensive national field trials, the U.S. Environmental Protection Agency approved the MUG-based methods for regulatory compliance. The 19th edition of *Standard Methods for the Examination of Water and Wastewater* will include the chromogenic (defined) substrate method for *E. coli*.

There are several reasons for choosing *E. coli* as the principal bacterial indicator in drinking water, wastewater, recreational waters, and shellfish waters. Research (Iowa State College, 1921; Dufour, 1977) has shown that *E. coli* are universally present in the feces of warm-blooded animals at densities of 10^8 to 10^9 per gram, and comprise nearly 95 percent of the coliforms in feces (see Tables 19.2 and 19.3). Based on this data, it can be concluded that *E. coli* would <u>always</u> be present in any fecal contamination event, unlike *Klebsiella, Enterobacter,* and *Citrobacter* which may be present, but at much lower densities.

3 PUBLIC HEALTH DECISIONS

By understanding which genera the indicator systems enumerate, better and quicker public health decisions can be made in confidence. By using the most specific indicator, water utilities will know when there is truly a public health problem, and consumers will not be unnecessarily alarmed.

It is inappropriate to consider public advisories based on total or fecal coliforms unless *E. coli* has been identified. On the other hand, the presence of *E. coli* should trigger immediate

Table 19.2 *Relative Number of Fecal and Non-fecal Types of Coliform Bacteria in Various Substances. Source: Iowa State College Engineering Experiment Station Bulletin, 62 (1921), p.79.*

Sources	No. of Strains Observed	Percentage of Strains of *Enterobacter aerogenes* type	Percentage of Strains of *E.* coli type
Human feces	2534	5.9	94.1
Animal feces	1832	7.4	92.6
Water	2137	35.2	64.8
Milk	1382	43.1	56.9
Grain	288	81.7	18.3
Soil	853	88.1	11.9

Table 19.3 *Percentage of Genera of Coliforms in Human and Animal Feces.Source: Escherichia coli: The Fecal Coliform. A.P. Dufour. Special Technical Publication 65, American Society for Testing and Materials, pg 48-58, 1977.*

Animal (# examined)	*E. coli*	*Klebsiella* spp	*Enterobacter/ Citrobacter*
Chicken (11)	90	1	9
Cow (15)	99.9	--	0.1
Sheep (10)	97	--	3
Goat (8)	92	8	--
Pig (15)	83.5	6.8	9.7
Dog (7)	91	--	--
Cat (7)	100	--	--
Horse (3)	100	--	--
Human (26)	96.8	1.5	1.7
Average %	94.5		

public notification if repeat samples are positive for this organism, regardless of the densities found. Other studies have shown that presence/absence data, rather than numbers of bacteria per sample volume, is a more valid approach for water quality monitoring to assess health risk.

The ramifications of advising the consumer to boil water are very onerous, and such advice must be fully warranted as a real and immediate health risk. The ramifications include countless consumer inquiries to the water utility and health agencies, calls to the 911 switchboard that interfere with more urgent emergencies, exhausted supplies of bottled water, purchasing of point- of-use water treatment devices, and hardships affecting dental, medical, and food services. Any public advisory should be weighed in the light of the disruptive consequences that result from this action.

Although it is now possible to assay for the definitive indicator of fecal contamination, *E. coli*, many regulatory/health agencies and water analysts continue to use the traditional, but less specific coliform and fecal coliform groups. In reality, the use of MUG-based methods to detect *E. coli* provides the drinking water community and consumers with an analytical tool that signals real and recent contamination. Regulatory agencies need to examine their current policies and regulations that pre-date the development of *E. coli* specific (MUG-based) methods and are based on the less specific coliform and fecal coliform indicator systems.

4 RECOMMENDATIONS

Based on the availability and reliability of MUG-based methods to provide timely and specific information on the microbiological quality of drinking water, the following recommendations should be considered:

1. Water utilities should use the coliform group to monitor for treatment efficiencies and for general water quality within distribution and storage systems.
2. Fecal coliform method is not recommended for monitoring of fecal contamination.
3. *E. coli* should be the principal indicator used to monitor for fecal contamination.
4. MUG-based tests should be used exclusively for the rapid analysis of *E. coli*.
5. Public advisories should only be made when *E. coli* has been found.
6. Regulations that specify acceptable percentages of total coliform-positive samples should be reconsidered.
7. Regulations that reference coliform densities may not be appropriate.
8. Regulations that call for public advisories based on coliforms or fecal coliforms may not be appropriate.
9. Water utilities with chronic or seasonal coliform positive samples should determine the reasons and take appropriate actions to correct this problem.

5 CONCLUSIONS

The drinking water community, including the regulatory and health agencies, is responsible for ensuring that consumers are provided with pathogen-free drinking water. This responsibility implies the use of the best monitoring methods for assessing water quality. Regulations and policies that restrict the use of MUG-based bacteriological methods, use the less specific indicators (coliforms, fecal coliforms), or specify acceptable percentages of positive samples/ coliform densities compromise the ability to carry out this responsibility.

As water treatment practices change to minimize the formation of disinfection by-products, greater numbers of coliforms and fecal coliforms may be found in drinking water. This does not imply a higher probability that pathogens may be present. Unless current regulations are changed to reflect the use of these new bacteriological methods, the drinking water community and consumers will witness the issuance of more health advisories that are not science-based and erode consumer confidence.

References

American Public Health Association, *Standard Methods for Water Analysis*, Report of Committee, 1905.
AWWA Research Foundation. *Bacterial Regrowth in Distribution Systems*. 90532. 1987.
AWWA Research Foundation. *Assessing and Controlling Bacterial Regrowth in Distribution Systems*. 90567. 1989.
AWWA Research Foundation. *Comparison of the Colilert Method and Standard Fecal Coliform Methods*. 90647. 1994.
Bagley S.T., R.J. Seidler, H.W. Talbot, and J.E. Morrow. Isolation of *Klebsiella* from Within

Living Wood. *Appl. Environ. Microbiol.,* 1978; 36(1):178-185.

Caplenas N.R., and M.S. Kanarek. Thermotolerant Nonfecal Source of *Klebsiella pneumoniae*: Validity of the Fecal Coliform Test in Recreational Waters. *Am. J. Public Health*, 1984; 74(11):1273-1275.

Covert T.C., E.W. Rice, S.A. Johnson, D. Berman, C.H. Johnson, and P.J. Mason. Comparing Defined-Substrate Coliform Tests for the Detection of *Escherichia coli* in Water. *Jour. AWWA*, 1992; 84:98-104.

Dufour P. *Eschrichia coli:* The Fecal Coliform. Special Technical Publication 65, *Amer. Soc. Testing and Materials,* 1977; 48-58.

Duncan I.B.R. Waterborne Klebisella and Human Diseases. *Toxicity Assessment-An International Journal,* 1988; 3(5):581-598.

Edberg S.C., M.J. Allen, D.B. Smith, and The National Collaborative Study. National Field Evaluation of a Defined Substrate Method for the Simultaneous Enumeration of Total Coliforms and *Escherichia coli* From Drinking Water: Comparison With the Standard Multiple Tube Fermentation Method. *Appl. Environ. Microbiol.,* 1988; 54(6):1595-1601.

Edberg S.C., M.J. Allen, D.B. Smith, and The National Collaborative Study. National Field Evaluation of a Defined Substrate Method for the Simultaneous Enumeration of Total Coliforms and *Escherichia coli* From Drinking Water: Comparison With Presence-Absence Techniques. *Appl. Environ. Microbiol.,* 1989; 55(4):1003-1008.

Edberg S.C., J.E. Patterson, D.B. Smith. Differentiation of Distribution Systems, Source Water, and Clinical Coliforms by DNA Analysis. *J. Clin. Microbiol.,* 1994; 32:139-142.

Iowa State College Engineering Experiment Station, Bulletin 62, 1921; pg. 79.

Rice E.W., M.J. Allen, and S.C. Edberg. Efficacy of β-glucuronidase Assay for the Identification of *Escherichia coli* by the Defined Substrate Technology. *Appl. Environ. Microbiol.,* 1990; 56:1203-1205.

World Health Organization, *Guidelines for Drinking-water Quality.* 1993; Volume 1, Recommendations.

* The views expressed are those of the author and do not necessarily reflect those of the AWWA Research Foundation.

Chapter 20
Viable But Injured Coliforms In Drinking Water Supplies

J.L. CRADDOCK, FIMLS, and R.G. CASTLE, BSc, MSc, (Member)

Microbiologist and Public Health Scientist respectively, North West Water Ltd

1 INTRODUCTION

This paper reports on an investigation into the occurrence and origin of injured coliforms in four drinking water sources in the North West of England, and how filter operation may contribute to their presence.

The technique used to identify injured coliforms was developed in America in the early eighties (LeChevallier et al., 1983; LeChevallier and McFeters, 1985), using principles established in food microbiology to resuscitate faecal indicators (Mudge and Smith, 1935; Heinmets et al., 1954). LeChevallier et al., produced selective medium called m-T7 (membrane tergitol 7) agar to show the presence of coliforms damaged, but not killed, by water treatment processes. Traditional bacteriological tests can fail to detect injured bacteria because some culture media components are toxic (McFeters 1982; McFeters 1991). However, if these bacteria are viable there may be implications for public health and regulatory control (LeChevallier and McFeters, 1985; McFeters, 1991, McKay, 1992).

Most work on injured coliforms in drinking water has concentrated on American water supplies (LeChevallier et al., 1983; LeChevallier and McFeters, 1985; McFeters et al., 1986; Clark 1988; McFeters 1991). In the first part of our study, the frequency of injured coliforms in four drinking water supplies in the North West of England, was examined by comparing counts of total coliforms on m-T7 agar with those detected by the standard UK method using membrane lauryl sulphate broth (Anon 1994).

The second part of the study investigated breakthrough of coliform bacteria immediately following backwashing of a rapid sand filter and the potential for this to overwhelm the disinfection process. Such breakthrough has been described previously (Bucklin and McFeters, 1991; Denny and Broberg, 1992).

2 METHODS

2.1 Study sites

Four treatment plants were chosen for their bacteriological performance (Table 20.1), each having different treatment processes. Treated water from plant A rarely failed bacteriological tests and was effectively a control site. The other sites had a history of occasional bacteriological failure. Plant D served a large distribution network in which

Table 20.1 *Study Sites*

Site	Water source	Treatment Details
Treatment plant A	Borehole of high quality	Aeration, chlorination & pressure filtration to remove manganese
Treatment plant B	Partially treated surface water from an aqueduct	Chlorination & pH correction
Treatment plant C	Impounded upland water	Coagulation with alum, rapid filtration, chlorination & pH correction
Treatment plant D	Class 1 river water without bank - side storage	Screening, coagulation with alum, sedimentation, rapid filtration through sand with activated carbon, pH correction & chlorination
Distribution system	Treatment plant D	As D but with secondary chlorination

bacteriological failures occurred regularly.

2.2 Sample collection

Company Water Quality Officers sampled drinking water from metal sampling taps on treatment plants, and from customer premises in distribution. These were collected in accordance with standard methods (Anon 1994) and analysed within 6 hours.

Samples of filtered, but unchlorinated water, were collected before, during, and after backwashing one of 16 rapid filters at treatment plant D. Further samples were taken at the treatment plant outlet after two hours, estimated from local knowledge, to cover the time of travel of water from the filter to the plant outlet.

2.3 Bacteriological analysis

Two commercial media were used: Membrane Lauryl Sulphate Broth (MLS, DIFCO 1841-17) for routine tests and Membrane Tergitol 7 agar (m-T7, DIFCO 0018-15-3) for recovery of injured coliforms. Batches were prepared weekly according to manufacturer's instructions and tested for sterility and ability to support growth before use.

Two aliquots of 100 ml were taken from each sample and filtered through 0.45μm sterile membranes (Gelman, 60068) using standard methods for membrane filtration (Anon 1994). The order of inoculation of each medium was alternated daily to remove experimental bias. Membranes were incubated at 30°C for 4 hours then 37°C for 14 hours.

Smooth yellow convex colonies on m-T7 or MLS were counted as presumptive coliforms (LeChevallier *et al.*, 1983). All isolates were subcultured onto MacConkey agar before being identified by the API system, and onto yeast extract agar for oxidase testing. Confirmed coliforms were oxidase negative and produced acid in lactose peptone water.

2.4 Statistical analysis

Any differences in the frequency of coliform counts between the two media was tested for significance by using the binomial theorem. Samples were assumed to contain no coliforms where analysis using m-T7 media had indicated their absence.

3 RESULTS

In all, 247 samples of treated water were analysed, of which 173 were final waters from water treatment plant outlets and 74 were from distribution. A further 28 were taken from a partially treated water source.

Results (Table 20.2) show an average of 10.4% of all the samples taken from the treatment plant outlets contained injured coliforms. These were mainly anaerogenic coliforms. Only 0.6% of the samples failed routine bacteriological tests. Samples taken from the distribution system did not contain significant numbers of injured coliforms. Only 2 (2.7%) out of 74 samples contained injured coliforms. None failed routine tests using MLS media.

Comparison of the numbers of coliforms and injured coliforms in partially treated water from treatment plant B showed that of the 92.9% of samples identified as containing injured coliforms, nearly all (89%) would have been detected by the standard coliform test.

3.1 Rapid filter study

Large numbers of coliform bacteria were identified in the samples taken immediately after back-washing. The numbers gradually declined over the next few hours to a small

Table 20.2 *Frequency of coliforms isolated using two different bacteriological media*

	SAMPLE LOCATION						
	Treatment plant A outlet	Treatment plant B		Treatment plant C outlet	Two aqueducts on outlet of treatment plant D		Distribution system from Treatment plant D
		Partially treated	Final		Aqueduct 1	Aqueduct 2	
Total number of samples analysed	25	28	32	32	42	42	74
Number of samples containing coliforms (Tested with MLS media)	0 (0%)	25 (89%)	0	0	1 (2.4%)	0	0
Number of samples containing injured coliforms (Tested with m-T7 media)	0 (0%)	26 (92.9%) No API ID carried out	4 (12.5%) No API ID carried out	4 (12.5%) No API ID carried out	4 (9.5%) (One identified as *Klebsiella spp.*)	6 (14.3%) (Two identified as *Klebsiella spp.*)	2 (2.7%) (*Enterobacter spp & Klebsiella spp.*)
Statistical significance of the difference in results between MLS and m-T7 media	Not significant	Not significant	P < 0.06	P < 0.10	P <0.16	P < 0.02	Not significant

fraction of that initially detected. The results of analysis using m-T7 media indicate about 25% of the coliforms were injured.

A peak of presumptive injured coliforms occurred in the final chlorinated water, at a time when the first backwash water was expected to have reached the plant outlet. The full results of the rapid gravity filter study are shown graphically in Figures 20.1 and 20.2.

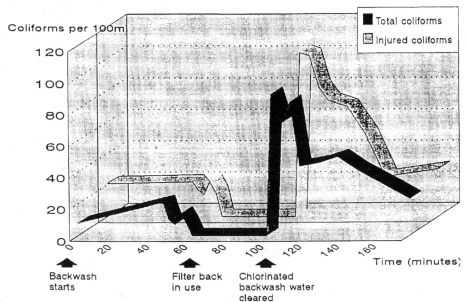

Figure 20.1 *Coliforms detected during a backwash cycle*

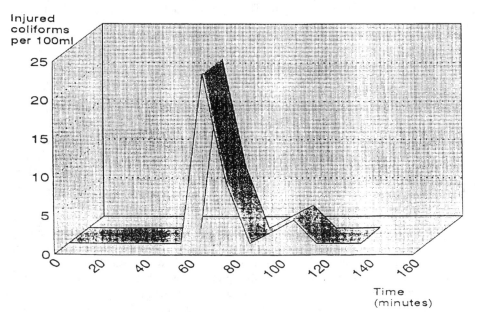

Figure 20.2 *Presumptive coliforms in the final water*

4 GENERAL OBSERVATIONS

Most injured coliform isolates were anaerogenic.

The colonies on m-T7 agar were significantly smaller and less yellow than those on MLS broth. This made it more difficult to count colonies on m-T7 media.

API tests identified three isolates from the final water of treatment plant D as the coliform organisms *Klebsiella rhinoscleromatis*, and two isolates from the distribution system, as *Enterobacter cloacae* and *Klebsiella oxytoca*. **All other isolates tested by API were o-nitrophenyl-β-D-galactopyranoside (ONPG) negative**. These could not be identified with confidence using API tests. This is an obvious anomaly because these organisms must have possessed the ß-galactosidase enzyme in order to ferment lactose.

5 DISCUSSION

Injured coliforms were detected in samples of final water which passed routine tests, but it is not clear from the literature if the presence of these bacteria in the final water is a health concern (LeChevallier and McFeters, 1985; McFeters 1991; McKay 1992).   There is no evidence injured coliforms cause significant contamination of the distribution system examined. This may be because of secondary chlorination or because coliforms find the environment within the water mains too hostile to survive. The tests using m-T7 media appear more sensitive, perhaps by an order of magnitude for some waters.

It is difficult to compare the results of this study with similar studies in America because MLS media, used for comparison with m-T7 in this study, is more efficient at detecting coliform bacteria than m-Endo media, used in American studies (Anon 1980).

Some bacteria may be injured prior to water treatment (Bissonnette *et al.,*   1992), although the filter study indicates this is less than 25%. No injured coliforms were detected in the treated water from the borehole. This probably reflects the high quality of the raw water in the aquifer.

5.1 Filter Study

The investigations show that immediately following filter back-washing, bacteria are able to breakthrough. This is in line with the observations of other workers (Bucklin and McFeters, 1991; Denny and Broberg, 1992), although the proportion of injured coliforms was less than measured by Bucklin *et al.,*. Coliforms probably breakthrough at the start of the filter run because the filter has not settled and because the filter pores are free from material which accumulates within the sand over the course of a filter run. Some bacteria breaking through may be contained within the matrix of waterborne particles and thereby protected from injury (Ridgeway and Olson, 1992). No bacteria were found in samples of water taken while back-washing was in progress because this water was chlorinated.

Breakthrough of bacteria following back-washing, may partially overwhelm the disinfection process and may explain why injured coliforms were detected in some treated waters. The output of individual filters is considerably diluted at large treatment plants, such as plant D. This may help prevent the disinfection process being overwhelmed by a short period of breakthrough from each filter. Most probably, the effects are greater where filters are not brought back into operation by "slow start up" procedures.

It is difficult to explain why many of the bacteria identified as injured coliforms could not be identified further by API tests. The medium may not be sufficiently selective, injuries may affect the organism's API profile, or the API system may not be suitable because of its emphasis on clinical isolates. However, it is of concern many of the organisms identified appeared ONPG negative on API testing. Under existing procedures, these bacteria are identified as coliforms because of their ability to ferment lactose.

Further investigations are required. There is a need to establish the extent to which injured bacteria pass through modern treatment processes and enter distribution systems. It is important to know if injured bacteria recover in distribution systems and indicate a health risk, and whether they are then detected by routine tests. Potentially, tests for injured coliforms provide a means of monitoring improvements in disinfection processes, and this requires investigating. Finally, the tests themselves need to be improved to enable full identification of the bacteria, perhaps by resuscitation in nutrient broth before identification.

6 CONCLUSION

Injured coliforms, as classified by this study, were detected in the final treated water from all but the borehole supply. This indicates that while some water treatment plants meet present standards for drinking water quality, they may not operate with a high margin of safety.

The treated water in the distribution system supplied by treatment plant D did not contain significant numbers of injured coliforms. This may be the result of new secondary chlorination systems at several points in this distribution system.

The m-T7 tests for injured coliforms are more sensitive than the routine MLS tests. Looking for injured coliforms therefore enables the plant operator to experiment with a disinfection process and optimise it. This can presumably be achieved with a safety margin, measurable in terms of the frequency of detection of injured coliforms and without exceeding existing regulations.

Coliform bacteria, some of which are injured, may break through a filter in the first hour following back-washing when the water is more turbid. The dilution effect of combining the output of a large number of filters at different stages of the backwash cycle probably prevents overloading of the disinfection process. However, breakthrough might explain the intermittent occurrence of coliforms in treated water.

An anomaly in the results requiring further investigation is that all the bacteria tested positively for coliforms when cultured in lactose peptone water, but many could not be identified by API testing. The reasons for this are not understood, but one possibility is that their injuries affect the API tests.

The study of injured coliforms is one of many tools which may help identify inadequacies in treatment processes, otherwise overlooked when using routine bacteriological tests.

Acknowledgements

The Authors wish to thank Ken Oliver, Dennis Blake, Ralph Albinson, John Daykin and all the other laboratory staff at Warrington, Denton and Sutton Hall laboratories for carrying out the work. Paul Burke is thanked for his help with the statistics. Paul West is thanked for his assistance in editing.

References

Anon (1980) Joint Committee of the Public Health Laboratory Service & Standing Committee of Analysts. A comparison between minerals modified glutamate medium and Lauryl tryptone lactose broth for the enumeration of coliform organisms and Escherichia coli in water: comparison of Tergitol 7 and Lauryl sulphate with Teepol 610. Journal of Hygiene, **85**, 181-91.

Anon (1994) The Microbiology of Drinking Water 1994. Part 1 - Drinking Water. London HMSO.

Bissonnette, G.K., Jezeski, J.T., Thomson, C.A., and Stuart, D.G. (1974) Comparative survival of indicator bacteria and enteric pathogens in well water. Applied Microbiology, **27**, 823-829.

Bucklin, K.E., and McFeters, G.A. (1991). Penetration of coliforms through municipal drinking water filters. Water Resources, 1991, **25**, 8, 1013-1017.

Clark, T.F. New culture medium detects stressed coliforms, Opflow, (1988), **11**, 3-6.

Denny, S., and Broberg, G. (1992). Progress report on the operation of rapid filters used in water treatment. Foundation for Water Research, Allen House, The Listons, Liston Road, Marlow, Bucks.

Heinmets, F., Taylor, W.W., and Lehman, J.J. (1954). The use of metabolites in restoration of the viability of heat and chemically inactivated Escherichia coli. Journal of Bacteriology, **67**.

LeChevallier, M.W., Cameron, S.C., and McFeters, G.A. (1983). New medium for improved recovery of coliform bacteria from drinking water. Applied and Environmental Microbiology, **45**, 484-492.

LeChevallier, M.W., and McFeters, G.A. (1985). Enumerating injured coliforms in drinking water. Journal of American Water Works Association, **77**, 81-87.

McFeters, G.A., Cameron, S.C., and LeChevallier, M.W. (1982). Influence of diluents, media, and membrane filters on the detection of injured coliform bacteria. Applied and Environmental Microbiology, **43**, 97-103.

McFeters, G.A., LeChevallier, M.W., and Kippin J.S. (1986). Injured coliforms in drinking water. Applied and Environmental Microbiology, **51**, 1, 1-5.

McFeters, G.A. (1991). Drinking water microbiology progress and recent developments. Springer-Verlag, New York, 478-492.

McKay, A.M. (1992). Viable but non-culturable forms of potentially pathogenic bacteria in water. Letters in Applied Microbiology, **14**, 129-135.

Mudge, C.S., and Smith, F.R. (1935). Relation of action of chlorine to bacterial death. American Journal of Public Health.

Ridgeway, H.F., and Olson, B.H. (1982). Chlorine resistance patterns of bacteria from two drinking water distribution systems. Applied and Environmental Microbiology, **44**, 4, 974-987.

Chapter 21
Does *Klebsiella Oxytoca* Grow In The Biofilm Of Water Distribution Systems?

Packer[1], PJ., Holt[2], DM., Colbourne[3], JS. and Keevil[1], CW.

[1] Microbial Technology Dept, CAMR, Salisbury, Wilts, SP4 0JG
[2] Thames Water Utilities, Manor Farm Road, Reading, Berks, RG21 0JN
[3] Thames Water Utilities, Walton Advanced Water Treatment Works, Hirst Rd, Walton-on-Thames, Surrey, KT12 2EG

1 INTRODUCTION

Potable drinking water has diverse microbial flora and it has been universally shown that bacteria can colonise the surfaces of a water distribution system forming a biofilm (Allen *et al.*, 1980, Ridgeway and Olson, 1981). Most reports describe the biofilms as a structure built up of different micro-organisms (Costerton *et al.*, 1986; Keevil *et al.*, 1992 and Ridgeway and Olson, 1981). The microcolonies form stacks upon each other and are surrounded by water channels, extracellular polymers and minerals from the immediate environment (Rogers, *et al.*; Keevil *et al.*, 1993).

The occurrence of coliforms in drinking water is of major interest to the water industry. Coliforms can enter distribution systems in many ways including back siphonage, cross connections, line breaks and repairs, inadequate or partial treatment, breakthrough or particle associated coliforms and survival and repair of injured coliforms (McFetters *et al.*, 1984). Growth or survival of coliforms in distribution systems is well documented (Olson and Nagy, 1984; Colbourne, 1985; Colbourne *et al.*, 1988; DWI Annual Report 1990; Mackerness *et al.*, 1991; Colbourne *et al.*, 1991; Mackerness *et al.*, 1993; Block, 1995). One possible mechanism for microbial survival may be the ineffectual penetration of disinfectants through the biofilm matrix, thus rendering any bacteria resistant to disinfection. It has been shown experimentally that coliforms can colonise and grow within biofilms on bitumen coated tiles simulating pipe surfaces (Mackerness *et al.*, 1993; Holt and Packer, 1995). It is probable that pieces of biofilm slough off from the walls of pipes releasing coliforms into drinking water (Mackerness, *et al*, 1993).

This study was designed to assess the effect of six waters, with different chemistry, on the ability of a potable water heterotrophic bacterial population supplemented by an environmental strain of *K. oxytoca* to colonise surfaces exposed in a chemostat model of a water distribution system.

2 MATERIALS AND METHODS

2.1 Test Waters

Three of the waters were taken from different sites within the London area, two from Water Treatment Works, Coppermills WTW and Kempton WTW, and one at the distal part of a distribution system. The fourth and fifth waters were Distribution water modified by the addition of ammonia and a sodium acetate buffer (LeChevallier *et al*, 1993) respectively to increase the nutrient content. The final water was an Advanced Treated Water (Kempton AWT) that had been subjected to ozonation and granulated activated carbon treatment. Ozonation breaks down large molecules leaving smaller carbon based molecules that may be more assimilable by bacteria.

2.2 Model System

A chemostat model was designed similar to that previously described by Mackerness *et al.*, (1993). A seed vessel was maintained at 15°C and inoculated with a mixed environmental inoculum of heterotrophic bacteria (Mackerness *et al.*, 1993). The vessel was then fed with sterile Distribution water (filtered through 0.2u, Pall filters) at a flow rate of 25 ml/hr. A fifth of the effluent from this vessel fed the test vessel for a period of two weeks.

During this two week period the test vessel, which was maintained at 20°C and contained 1 cm² bitumen coated steel tiles suspended by titanium wire, was fed with sterile Distribution water at a flow rate of 95 ml/hr. After the two week period a 24 hr old pure culture of *K. oxytoca* grown on R2A medium was then scraped from the agar surface and inoculated into the test vessel and left in batch conditions for 30 minutes. The seed was terminated and the chemostat was then continually fed with the sterile test water at a flow rate of 100 ml/hr over the test period of 21 days. Tiles were removed at intervals for the determination of biofilm bacteriology and samples of the liquid phase of the chemostat were taken at similar times for bacterial analysis. *K. oxytoca* counts were determined using CLED agar grown at 37°C for 24 hr. Total heterotrophic counts were enumerated on R2A incubated for 7 days at 22°C.

3 RESULTS

Total heterotrophic numbers were not statistically different in the biofilm of any of the test waters over the 21 day test period (Figure 21.1) although initial colonisation rates appeared to show several fold differences. The different composition of the water had no apparent effect on total culturable numbers. Similar observations were seen in the planktonic phases (Figure 21. 2). The fact that the numbers remained relatively constant in all of the test waters suggest

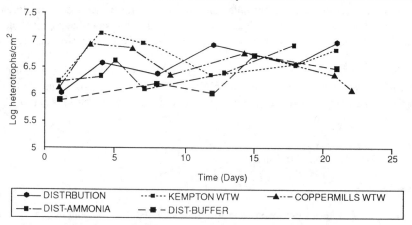

Figure 21.1 *Log total heterotrophs in the biofilm phase in each of the vessels*

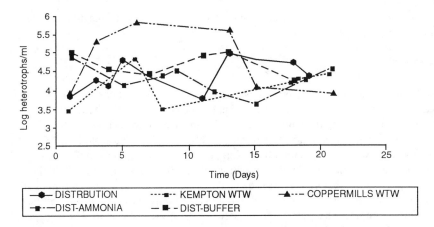

Figure 21.2 *Log total heterotrophic counts in the planktonic phase in each of the test waters*

that the organisms are growing to a similar degree. There is a slight indication that heterotrophs grow better in the Coppermills WTW but this difference is not statistically significant suggesting that these values reflect natural variation.

Similar patterns of depletion of *K. oxytoca* numbers in the planktonic phase were observed in each of the test waters (Figure 21.3). In general, a gradual depletion to very low numbers of *K. oxytoca* was observed over a two week period. For Kempton WTW depletion took 3 weeks to reach a point at which bacterial numbers were on or below the limit of detection. *K. oxytoca* numbers in Distribution water appeared to plateau out in the final week of the run. Addition of ammonia or sodium acetate buffer had no effect on coliform survival since none of these curves were statistically different.

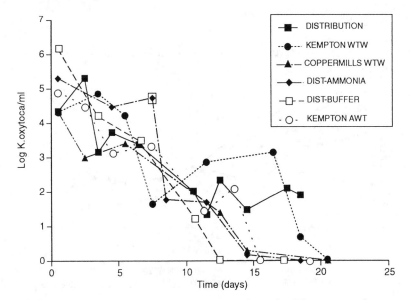

Figure 21.3 *Log K. oxytoca in the planktonic phase in each of the test vessels*

Chemostat theory for a conservative parameter in water shows that the dilution rate (D) follows an exponential relationship:-

Nt=Noe^{-Dt}

and for this model D= 0.2 h^{-1}. Theoretical washout under these conditions indicates that if *K. oxytoca* was simply washing out then it would not be detectable after 72 hours. It is obvious that *K. oxytoca* is surviving for a much longer period in all the test waters suggesting that it is growing slowly in the planktonic phase of the model. The calculated washouts, D (h^{-1}), for each of the waters were:- Distribution-0.03, Kempton WTW-0.013, Coppermills WTW-0.02, Distribution Water + ammonia-0.03, Distribution water-0.03 and Kempton AWT -0.03.

Figure 21.4 shows that *K. oxytoca* is depleted to a similar degree in the biofilm of all of the test waters, not being influenced by different water composition. At the end of the test period coliform numbers had decreased by between 93 to greater than 99%, but were still detectable. This suggests that the *K. oxytoca* was surviving in the biofilm and leads to the question as to whether it is growing. From these results it is not clear. If it is growing then its is doing so very slowly but there is no conclusive evidence for this.

There are difficulties in mathematically modelling the growth of *K. oxytoca* in the chemostat system. Various factors influence its presence in the planktonic and biofilm phase which are not taken into account in conventional chemostat theory. These are related to the biofilm attached to the side of the glass vessel and include, the survival and generation time of *K. oxytoca*, and the detachment and attachment rate of the coliform to and from the biofilm. It is therefore difficult to model the contribution of *K. oxytoca* from the biofilm on numbers in the planktonic phase with any great accuracy. Calculations of the maximum numbers of *K. oxytoca* that were available to enter the liquid phase from the biofilm were made using the formula:-

$$\frac{\text{Area x Number}}{\text{Vol over 24 h.}}$$

Where area is the chemostat surface on which the biofilm accumulates (235 cm^2), Number is

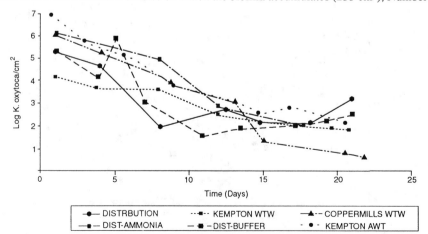

Figure 21.4 *Log K. oxytoca in the biofilm phase phase in each of the test vessels*

the number of bacteria detected on the bitumen coated tile (cm^2) and volume is the amount of liquid provided to the chemostat in 24 hours (2280 ml). After day 8 the numbers of *K. oxytoca* available from the biofilm are very similar to those detected in 1 ml of the planktonic phase, while on days 1 and 4 there are many more present in the planktonic phase. When the chemostat was originally seeded with *K. oxytoca* the organisms may randomly and loosely attach to the surface of the biofilm and therefore many of the organisms may simply be resuspended into the planktonic phase as they are not rigidly incorporated into the matrix. This could explain the initially large decreases in *K. oxytoca* in the biofilm. Even so there may be considerable reattachment to the biofilm throughout the experimental period as well as growth and death of the bacteria. The longer the *K. oxytoca* are in the biofilm the more chance they have of becoming incorporated and fixed, potentially forming colonies in distinctive areas. If pieces of the biofilm then slough off any colonies present will be released into the planktonic phase. As not all biofilm will contain *K. oxytoca* then its presence in very low numbers may allude detection by culture.

4 DISCUSSION

Other than the fact that *K. oxytoca* appeared to survive in the planktonic phase in Distribution water for a longer period of time than in the other test waters, there were no apparent significant differences in the detection of the bacteria in any of the waters. Nutrient loading either by the addition of ammonia, a sodium acetate buffer or potentially as a result of ozonation had no effect on the growth of the heterotrophic population or *K. oxytoca*.

Although *K. oxytoca* present in the planktonic phase washed out of the model in 21 days it survived longer than would be expected from the theoretical washout of 72 hr. This suggests that the bacteria are growing. However, given the constraints of the model relating to coliforms attaching to the walls of the chemostat this is by no means definite and many of these coliforms may originate from the biofilm. Numbers in the biofilm also decreased, and after 21 days, numbers were between 93 and greater than 99% lower than those at the beginning of the experimental period. *K. oxytoca* was still being detected in moderate numbers suggesting that it is surviving and possibly growing slowly in the biofilm. Any bacteria present in the biofilm would have two survival advantages. Firstly, they have increased access to trace minerals and organic nutrients from secondary metabolites of the biofilm consortium, and secondly, they may be protected from chemical disinfectants.

Increasing evidence suggests that coliforms can grow in distribution systems under a variety of conditions (Olson and Nagy, 1984; Colbourne, 1985; Colbourne *et al.*, 1988; Mackerness *et al.*, 1991; Colbourne *et al.*, 1991; Mackerness *et al.*, 1993; Block, 1995). Even if our results only support limited growth of *K. oxytoca* in biofilms they cast doubt on the use of coliforms as a marker of faecal contamination of water, since detection of coliforms in water would not confirm if they were of faecal origin or environmental strains growing or surviving in the biofilm.

Acknowledgements

This work was funded by Thames Water Utilities Ltd.

The authors would like to thank Joanne Dickenson and Sue Hill for their excellent technical assistance.

References

Allen, M. J., Taylor, R.H., Geldreich, E.E. The occurrence of microorganisms in water mains crustations. J. Am. Water Works Assoc 1980; 72: 614-625.

Block, J.C. Survival and growth of *E. coli* in drinking water distribution systems. Proc.Coliforms and *E. coli*: problem or solution? 1995, Leeds.

Colbourne, J.S. Materials usage and their effects on the microbiological quality of water supplies. J. Appl. Bact. Symp. Supp. (1985); 775-795.

Colbourne, J.S., Trew, M.B. and Dennis, P.J. Treatment of water for aquatic bacterial growth studies. J. App. Bact. (1988); 65: 79-85.

Colbourne, J.S., Dennis, P.J., Keevil, C.W. and Mackerness, C.W. The operational impact of growth of coliforms in Londons Distribution System. (1991); Proc Water Quality Technology Conference, Orlando, American Water Works Association, Denver, 799-810.

Drinking Water Inspectorate Annual Report 1990. Drinking water 1990-A report by the Chief Inspector. DWI, HMSO, London.

Costerton, J.W., Nickel, J.C. and Ladd, T.I. Suitable methods for the comparative study of free-living, and surface-asociated bacterial populations. In Poindexter, J.S and Leadbetter, E.R. (ed.), Bacteria in nature- Methods and special applications in bacterial ecology. 1986; 2: 49-84, Plenum Press, New York.

Holt, D.M. and Packer, P.J. Unpublished results,1995.

LeChevallier, M.W., Shaw, N.E., Kaplan, L.A. and Bott, T.L. Development of a rapid Assimilable Organic Carbon method for water. Appl. Environ. Micro. 1993, 59:1526-1531

Keevil, C.W., Walker, J.T. Normarski. DIC microscopy and image analysis of biofilms. Binary 1992; 4: 92-95.

Keevil, C.W., Dowsett, A.B. and Rogers, J. Legionella biofilms and their control. Society for Applied Bacteriology Technichal Series: Microbial Biofilm, 1993; 201-215.

Mackerness, C.W., Colbourne, J.S., Dennis, P.J., Rachwal, A. and Keevil, C.W. Formation and control of coliform biofilms in drinking water distribution systems. In Denyer, S.P., Gorman, S.P and Sussman, M. (ed.) Society for Applied Bacteriology Technical Series 1993; 30: 217-226, Blackwell Scientific, Oxford.

Mackerness, C.W., Colbourne, J.S. and Keevil, C.W. Growth of Aeromonas hydrophila and Escherichia coli in a distribution system biofilm model. In Morris, E.d., Alexander, L.M., Wyn-Jones, P. and Sethwood, J. (Eds), Proc. UK. Symp. Health Related Water Micro. (1991); 131-138.

McFetters, G.A., LeChevallier, M.W. and Domek, M. Injury and recovery of coliform bacteria in drinking water. EPA-600/2-84-166. 1984; US Environmental Protection Agency, Cincinnati, Ohio.

Olson, B.H. and Nagy, L.A. Microbiology of potable water. Adv. Appl. Microbiol. 1984; 30: 73-132

Ridgeway, H.F. and Olson, B.H. Scanning electron microscope evidence for the colonisation of a drinking water distribution system. Appl. Environ, Microbiol. 1981; 41: 274-287.

Rogers, J., Lee, J.V., Dennis, P.J. and Keevil, C.W. Continuous culture biofilm model for the survival of Legionella pneumophila and associated protozoa in potable water systems. In Health related water microbiolgy (Morris, R., Alexander, L.M., Wyn-Jones, P. and Sellwood, J. (Eds)) IAWPRC, London, 192-200.

Chapter 22
Multiplication Of Coliforms At Very Low Concentrations Of Substrates In Tap Water

D. van der Kooij

Kiwa N.V. Research and Consultancy, P.O. Box 1072, 3430 BB NIEUWEGEN, The Netherlands

1 INTRODUCTION

For nearly a century the absence of faecal and total coliforms has been an essential criterion for the microbiological safety of drinking water. Recent developments have shown that in certain situations pathogenic protozoans may penetrate the treatment system and drinking water complying with the coliform criteria may cause waterborne diarrhoea. Still, the coliform criteria are maintained in the United States as well as in Europe. The faecal coliforms, e.g. the thermotolerant coliforms, and more precisely representatives of the genus *Escherichia coli*, which are present in the intestines of the warm-blooded animals and man, are indicating faecal contamination. The other coliforms may have a faecal origin but many of these organisms can also multiply in the environment. Their presence in drinking water distribution systems therefore either indicates situations in which these organisms enter the distribution system or conditions in which these organisms can grow in the distribution system. In the first situation faecal contamination cannot be excluded but is very limited when *E. coli* is not detectable whereas in the second situation no contamination with faecal material has taken place. Recontamination of drinking water is a potentially dangerous situation. Moreover, regrowth is also undesirable because it may lead to a deterioration of drinking water quality. Finally, certain coliforms are considered as opportunistic pathogens (Geldreich and Rice, 1987). Aiming at controlling the level of total coliforms in drinking water therefore is an important objective in water treatment and distribution.

Increases of numbers of coliforms in drinking water during distribution have been an issue for debate for more than 60 years. In 1930, Baylis reported that 14 of 48 water supply companies in the USA responding to a questionnaire had problems with bacterial aftergrowths, involving coliforms (Baylis, 1930a). In 1990, 29 utilities of 164 responding to a survey in the USA, had one or more episodes in the past 3 to 4 years with more than 5% coliform occurrences (Smith *et al.* 1990). In some of these systems, seasonal effects were observed, whereas in others the nature of the occurrence was more chronic. Other reports from the USA (Wierenga, 1985; LeChevallier *et al.* 1991) and also the UK (Drinking Water Inspectorate, 1993) show that coliform occurrences exceeding the 5% level is a wide spread phenomenon.

Explaining the cause of coliform-positive samples is an essential part of the discussion about the sanitary significance of coliform occurrences. Moreover, the causes for coliform occurrences should be elucidated to enable effective curative and preventive measures. Baylis (1930a) related the occurrence of coliforms with open reservoirs in distribution systems.

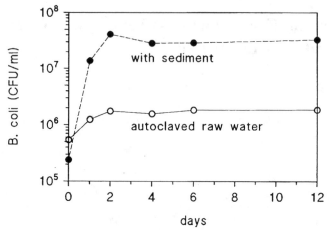

Figure 22.1 *Growth of B. coli in 400 ml of sterile raw water without and with 30 ml of autoclaved sediment (data from Baylis, 1930 b)*

Baylis (1930b) also observed that autoclaved decaying micro-organisms collected from chlorinated drinking water strongly enhanced the multiplication of coliforms in sterile water (Figure 22.1). He further demonstrated that coliforms appeared in samples collected from a pipe at flow rates high enough to pick up some accumulated sediment. From these and other observations he concluded that biomass accumulating on the pipe wall and in sediments is responsible for the multiplication of coliforms. Consequently, Baylis (1930b) stated that (i), drinking water should be free from suspended organic matter, (ii) be low in organic compounds that support bacterial growths; (iii), no open reservoirs should be connected to the distribution system and (iv) a chlorine residual should be maintained in drinking water during distribution.

Observations in practice and research in the past decade have supported the opinion of Baylis. The importance of sediments as growth substrate for coliforms has been confirmed by several investigators (Martin *et al.* 1982; Herson *et al.* 1991; Camper *et al.* 1991). Also specific materials in contact with drinking water e.g. leather (washers), jute yarn, lubricating oils, and wood have been shown to promote growth of coliforms (Burman and Colbourne, 1977; Seidler *et al.* 1977). Several investigators have demonstrated that coliforms in biofilms are protected from chlorine (Ridgway and Olson, 1982; Herson *et al.* 1987, LeChevallier *et al.* 1988). Factors affecting coliform regrowth in drinking water distribution systems have extensively been reviewed by LeChevallier (1990). A key issue in this review and in other recent studies is how to define water quality to prevent the formation of biofilms, which provide food sources for coliforms and also protect these bacteria against disinfectants. Several methods have been developed to determine the heterotrophic growth potential of drinking water. These methods are based on the argument that in most water types the availability of organic carbon to serve as growth substrate for micro-organisms is the growth limiting factor. These methods include: (i) assessment of the concentration of easily assimilable organic carbon (AOC) with growth measurements using a mixture of pure cultures (Van der Kooij,1990) and (ii), the determination of the concentration of biodegradable dissolved organic carbon (BDOC) (Servais *et al.* 1989, Joret and Lévy, 1986). These and other methods have been reviewed by Huck (1990). This paper describes (i), the occurrence of coliforms in drinking water distribution systems in the Netherlands and (ii) properties of coliforms to grow at very low substrate concentrations.

2 COLIFORM OCCURRENCES IN DRINKING WATER IN THE NETHERLANDS

Ground-water supplies, covering about two thirds of the total drinking-water supply, and most surface-water supplies in the Netherlands distribute drinking water without disinfectant residual. According to national legislation, faecal and other coliforms should not be present in samples (300 ml) of drinking water leaving the treatment facility. In 1993, 0.4% of all samples collected were coliform positive. In drinking water distribution systems, an average of 0.3% positive samples (100 ml) was observed (Versteegh *et al.* 1995). Examples of coliform observations are shown in Table 22.1 for a number of unchlorinated ground water supplies. Episodes of coliform occurrences were not observed in these supplies. In a survey on 20 distribution systems one or more coliforms were found in 0.7% of 411 samples of 300 ml volume (Van der Kooij, 1992).

Occasionally local coliform episodes are observed, e.g. due to heavy rainfall resulting in leaking wells, or the installation of a new main. In the latter case contamination of the main itself, or multiplication of coliforms on the lubricant used in the pipe joints has been observed. In these cases, the involved pipe section normally is not in operation. Other causes for coliform positive samples include: contaminated sampling points, multiplication of coliforms in a filter bed or multiplication in the distribution system. A strategy for further investigations and curative measures is available when coliforms are observed. When a sample appears to be positive in the presumptive test, additional samples are collected immediately to determine the extent and the cause of the presence of the coliforms in drinking water. Boiling orders usually are not given when no faecal coliforms are observed. In such situations flushing is used to improve water quality. When faecal contamination is observed, boiling orders are issued and measures including flushing and chlorination are used to remove the contamination (Nobel *et al.* 1995). Depending on the cause of the contamination, other measures such as closing part of a treatment facility may be appropriate in certain situations.

Coliform types isolated from drinking water are shown in Table 22.2. *Citrobacter freundii* is the predominant coliform in ground water supplies, but also *Enterobacter cloacae, Klebsiella pneumoniae* and *Klebsiella oxytoca* which predominate in surface-water supplies have been observed in ground water supplies. Representatives of the species of *Enterobacter* and *Klebsiella* has also been found in chlorinated supplies in the United States (Seidler *et al.* 1977; Martin *et al.* 1982; LeChevallier *et al.* 1987. The presence of *C. freundii* may be typical for unchlorinated ground water supplies.

Regrowth appears to be a major cause of the presence of coliforms in drinking water distribution systems. Such regrowth usually is observed in the summer months and may be

Table 22.1 *Frequencies of coliform-positive samples in 20 unchlorinated ground-water supplies*

Sampling location	Annual number of samples	Year		
		1991	1992	1993
Treatment facility (*)	1092	0.5	0.6	0.6
Reservoirs (*)	945	0.9	0.8	1.2
Dead ends (**)	2029	0.5	0.8	0.5

*, samples of 300 ml; **, samples of 100 ml.

occurring in certain parts of the distribution system (e.g. service reservoirs and dead end pipes). With heterotrophic plate counts on diluted-broth agar in unchlorinated drinking water ranging from several hundreds to several thousands CFU/ml, numbers of coliforms constitute a very small fraction (<0.001-0.01%) of the heterotrophic bacterial population in drinking water culturable on solid media (Van der Kooij, 1992). Obviously, coliforms are unable to compete effectively with the autochthonous heterotrophic bacteria for available substrates. Even the fluorescent pseudomonas and the aeromonas, which can grow at very low substrate concentrations (Van der Kooij *et al.* 1982; Van der Kooij and Hijnen, 1988) constitute only a minor fraction of the heterotrophic population, indicating the strong competitive capacity of the indigenous bacterial flora for available substrates. Still, in certain situations coliforms apparently multiply to such extent that their presence is observed in samples of 100 ml. In drinking water in the Netherlands, concentrations of easily available substrates (AOC) usually are in the range of 5 to 50 μg of acetate-C equivalents/l (Van der Kooij, 1992). The properties of coliforms to grow at low substrate conditions were studied as part of a continuing investigation aiming at (i), defining criteria for biologically-stable drinking water and (ii), obtaining information which may be used to define curative and preventive measures in distribution systems in addition to achieving biologically-stable drinking water in treatment.

3 GROWTH SUBSTRATES FOR COLIFORMS

Studies on the nutritional versatility of coliforms, including *E. coli* all demonstrate that carbohydrates and a number of polyalcohols, are the favourite sources of energy and carbon for these organisms (Krieg and Holt, 1983; Roberts *et al.*, 1992). In addition also a number of carboxylic acids and a few amino acids can serve these purposes. In a study on the chemotactic behaviour of *E. coli* it has been observed that some compounds including D-galactose, D-glucose, and maltose had a threshold concentration for a chemotactic response as low as a few μM (Adler *et al.* 1973). Low threshold values were also observed for the amino acids L-aspartate (0.01 μM), L-serine (0.1 μM) and L-cysteine (1 μM) (Hedblom and Adler, 1983). *E.*

Table 22.2 *Identity of coliforms observed in drinking water samples from a surface water supply and from ground water supplies (cf Table 22.1)*

Organism	SWS (*)	GWS (**)
Escherichia coli	8.4	11
Citrobacter freundii	7.7	33
Enterobacter cloacae	27.3	10
Enterobacter sakazakii	4.9	8
Enterbacter agglomerans	4.9	3
Enterobacter amnigenus	no	4
Enterobacter fergusonii	no	1
Klebsiella pneumoniae	10.5	4
Klebsiella oxytoca	25.9	3
Other species (***)	10.6	10

*, surface water supply; 1977-1982.
**, ground water supplies; 1991-1993.
***, including *Aeromonas* spp.
no, not observed

coli strains possess uptake systems for a wide range of compounds, including carbohydrates, amino acids, peptides and carboxylic acids (see below). Uptake of a compound does not necessarily mean that it can serve as an energy source. Uptake of amino acids and peptides enables the organism to utilize these compounds directly for biomass synthesis. Growth measurements with pure cultures of *P. fluorescens* and *A. hydrophila* revealed that these organisms utilize most if not all naturally occurring amino acids when present at low concentrations ($\leq$ 10 µg of C/l) in a substrate mixture, including those which do not serve as a sole source of energy and carbon (Van der Kooij *et al.* 1982; Van der Kooij and Hijnen, 1988). Many coliform bacteria may have a greater nutritional versatility than *E. coli*.

4 GROWTH KINETICS

The rate of growth of a bacterial species in relation with the concentration of the growth limiting substrate is given by the Monod equation:

$$V = V_{max} \times S/(K_s + S) \qquad (1),$$

where: V = growth rate (doublings/h) in the exponential phase, V_{max} is the maximum growth rate (h^{-1}), S is concentration of the growth-limiting substrate (mg/l) and K_s, the substrate saturation constant which equals the value for S at which $V = 1/2 \, V_{max}$. Determining the growth constants V_{max} and K_s enables calculations on the rate of growth of coliforms at very low concentrations of substrates.

For *E. coli*, growing on glucose, K_s-values of about 1 µM (72 µg C/l) have been reported (Von Meyenburg, 1971). Hyashi and Li (1965) found a K_s value of 0.9 µM (32 µg C/l) for glycerol. Camper *et al.* (1991) reported that K_s values of coliforms for yeast extract ranged from 50 to 240 µg/l.

The Monod equation also applies for the uptake of compounds. In this case K_t is the transport constant. K_t values of *E. coli* for some carbohydrates and amino acids are about 1 µM (Piperno and Oxender, 1968; Silhavy *et al.* 1978; Dahl and Manson, 1985). For dicarboxylic acids, K_t values of 15 to 30 µM have been reported (Kay and Kronberg, 1971, Lo *et al.* 1972). These values give an indication of the levels of substrate at which uptake and growth may become limited. Some uptake systems of coliform bacteria resemble those of *E. coli* (Dahl and Manson, 1985) but data seem to be scarce.

At very low substrate concentrations ($S << K_s$), equation 1 gives:

$$V = S \times V_{max}/K_s \qquad (2),$$

in which V_{max} and K_s are constants for a certain combination of organism-substrate and temperature and the quotient V_{max}/K_s is the substrate-affinity constant. Thus equation 2 indicates that the rate of growth is linearly related with the substrate concentration S, when $S << K_s$.

Bacteria use a part of the available growth substrate for maintaining the integrity of the cell. When substrate uptake does not meet the maintenance demands, the bacteria die from starvation. The die-off rate or endogenous decay rate (b, h^{-1}) can be considered as a negative growth rate. The substrate concentration below which no growth is possible for a certain organism-substrate combination therefore can be derived from:

$$V = V_{max} \times S/(K_s+S) - b \qquad (3)$$

S_{min} is the value for S at which $V = b$, hence:

$$S_{min} = b \times K_s/(V_{max}-b) \qquad (4)$$

Since $b << V_{max}$,

$$S_{min} = b \times K_s/V_{max} \qquad (5)$$

Values for the die-off rate b for aerobic bacteria utilizing organic compounds range from 0.01 to 0.03 h^{-1} (Servais *et al.* 1985).

5 MULTIPLICATION OF COLIFORMS AT LOW CONCENTRATIONS

Information about growth substrates and growth kinetics of coliforms other than *E. coli* appears to be very scarce. Investigations previously reported by the author (Van der Kooij and Hijnen, 1985) showed that coliforms isolated from drinking water supply systems can grow at relatively low concentrations of substrates. An experiment growing a *Klebsiella pneumoniae* strain in slow sand filtrate supplied with mixtures of substrates with each single compound at a concentration of 1 μg C/l revealed that only the mixtures of 18 carbohydrates and of 18 amino acids promoted growth (Van der Kooij and Hijnen, 1988). No growth was observed in slow sand filtrate without added substrates (blank) nor in the presence of the mixtures of carboxylic acids or the aromatic acids. From the yield of the organism on glucose (3 x 10^6 CFU/μg C), it was calculated that growth on the carbohydrates mixture (18 μg C/l) was equivalent to the utilisation of 11 μg of glucose-C/l, and the growth on the amino acids mixture (18 μg C/l) was equivalent with the utilisation of 3 μg glucose-C/l.

A total of 23 compounds which had been identified as sole source of carbon and energy for the *K. pneumoniae* isolate, were tested individually as growth substrates at a concentration of 25 μg of C/l. Also these tests were conducted at 15°C in pasteurized slow sand filtrate supplemented with ammonia-N as described previously (Van der Kooij and Hijnen, 1988). Figure 22.2 shows the growth rates (doublings/h) of the organism with the different compounds. With a number of carbohydrates, i.e. maltose, maltotetraose and arabinose, doubling times were below 10 h. With the exception of L-proline, the amino acids were very poor growth substrates at the concentration tested.

For a few compounds, the growth constants have been determined in tests with a range of substrate concentrations. The growth curves of a *K. pneumoniae* strain growing at a range of concentrations of maltopentaose are depicted in Fig. 22.3. This figure clearly demonstrates that even with this favourite substrate, growth is very slow (doubling time of about 30 h) at a concentration of 10 μg of C/l. Figure 22.4 shows the relationship between the maltopentaose concentration and the growth rate.

Assuming a decay rate of 0.01 h^{-1} for coliforms at 15°C enables an assessment of S_{min} values for the substrates shown in Figure 2 because the quotient of V/S equals the substrate affinity constant V_{max}/K_s (eq. 2). These hypothetical S_{min} values thus range from 1.7 μg C/l (for maltose) to 114 μg C/l (for alanine). Using higher or lower values for b will result in proportionally higher or lower S_{min} values.

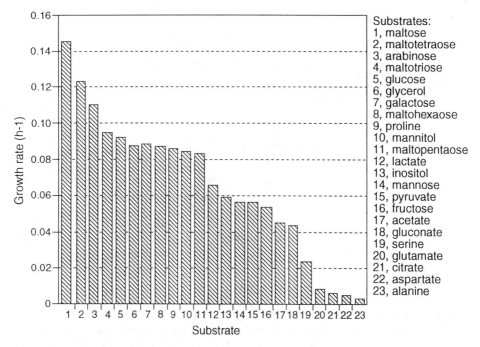

Figure 22.2 *Growth rates (doublings/h) of K. pneumoniae in slow sand filtrate supplemented with 25 μg of C/l of various substrates. Incubation temperature: 15 °C.*

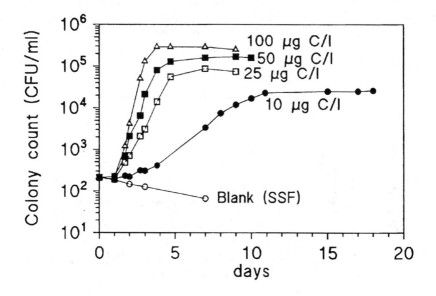

Figure 22.3 *Growth curves of K. pneumoniae in slow sand filtrate (SSF) at different concentrations of maltotetraose. Incubation temperature: 15 °C.*

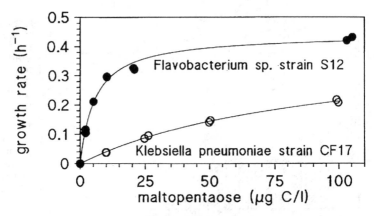

Figure 22.4 *Growth rates of K. pneumoniae and Flavobacterium species strain S12 at various concentrations of maltopentaose. Data for Flavobacterium sp. strain S12 derived from: Van der Kooij and Hijnen, 1985.*

6 COMPETITION FOR SUBSTRATES

The K_s values and substrate affinities of coliforms for carbohydrates are much higher than such constants observed in bacteria normally present in drinking water, i.e. *Flavobacterium* spp., *Aeromonas* spp. and *Pseudomonas* spp. The relationship between the growth rates of a *Flavobacterium* strain and the concentration of maltopentaose is also presented in Figure 22.4. This *Flavobacterium* strain multiplied much more rapidly at all the concentrations tested than the *K. pneumoniae* (Van der Kooij and Hijnen, 1985). Even at a concentration of 1 µg of C/l, which is below the S_{min} value for *K. pneumoniae*, the *Flavobacterium* strain would have a doubling time of about 12 h. Figure 22.5 shows the growth rates of an *A. hydrophila* strain and an *Enterobacter cloacae* in slow sand filtrate supplemented with various concentrations of soft soap. This soap was tested as a growth substrate, because observations in practice indicated that coliforms present in newly laid mains multiplied on lubricants in pipe joints.

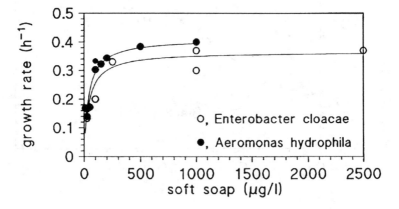

Figure 22.5 *Growth rates of Enterobacter cloacae and Aeromonas hydrophila on soft soap added to slow sand filtrate. Temperature: 25 °C.*

Table 22.3 *Growth kinetics of coliforms and a few other bacteria isolated from drinking water*

Organism	Substrate	Temp ('C)	K_s (µg C/l)	V_{max} (h-1)	$G_{S=1}$* (h)	S_{min} (µg C/l)	Reference
E. cloacae	glucose	15	60	0.21	285	2.8	Van der Kooij and Hijnen, 1985
C. freundii	glucose	15	95	0.17	558	5.5	Van der Kooij and Hijnem, 1985
K. pneumoniae	glucose	15	105	0.32	328	3.3	This report
A. hydrophila	glucose	15	16	0.28	57	0.6	Van der Kooij and Hijnen, 1988
K. pneumoniae	maltose	15	51	0.49	104	1.0	Van der Kooij and Hijnen, 1987
K. pneumoniae	maltopentaose	15	92	0.41	224	2.2	Van der Kooij and Hijnen, 1987
Flavobacterium sp.	maltopentaose	15	5.6	0.44	12.7	0.13	Van der Kooij and Hijnen, 1985
E. cloacae	Soft soap	25	43	0.36	119	1.2	This report
A. hydrophila	Soft soap	25	41	0.41	100	1.0	This report

*, $G_{S=1}$, the generation time (hours) at a concentration of 1 µg C/l,
  calculated from eq. 2 with S = 1 µg C/l and G = 1/V.

Soft soap obviously is very effective in promoting the multiplication of both organisms. A comparison of growth kinetics of coliforms and a few bacteria isolated from drinking water for a few substrates is given in Table 22.3. For all combinations of coliforms and substrates tested, the doubling time at a concentration of 1 µg of C/l exceeds 100 h. With an endogenous decay rate of 0.01 h⁻¹, multiplication would not be possible at this concentration. Consequently, bacteria normally occurring in drinking water e.g. *Flavobacterium* spp. and also *Aeromonas* spp. easily out compete the coliforms at substrate concentrations in the range of a few µg/l.

7 STARVATION SURVIVAL

In an experiment in which *K. pneumoniae* had been cultured in slow sand filtrate with 1 mg glucose-C/l, the numbers of the organism were determined during prolonged storage of the water samples at 15°C. The viable count of this organism clearly declined after about 30 days, but after nearly two log units reduction culturable bacteria survived for a long period of time (Figure 22.6). Other reports have also demonstrated that coliforms (including *E. coli*) can survive for a long period when present as a pure culture (Flint, 1987). The ability to survive may be important in explaining the presence of coliforms in drinking water distribution systems.

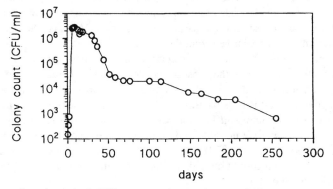

Figure 22.6 *Growth and survival of K. pneumoniae in slow sand filtrate supplemented with 1 mg of glucose-C/l. Temperature: 15°C.*

8 DISCUSSION AND CONCLUSIONS

Coliforms may be present in drinking water distribution systems as the result of multiplication. Carbohydrates are favourite growth substrates for coliforms, but information about substrate utilization kinetics for coliforms appears to be scarcely available. Data on substrate uptake kinetics of *E. coli* and multiplication at low concentrations suggest that substrate uptake and growth is limited at concentrations below 1 μM of potential growth substrates. Growth measurements conducted with pure cultures of a few coliforms confirmed these data, but also showed that coliforms can grow at substrate concentration as low as 10 μg of C/l. However, comparison of growth kinetics also revealed that bacteria normally occurring in drinking water multiply much faster on the substrates potentially available for coliforms. Such organisms even multiply at   concentrations below the threshold value (a few μg of C/l) for growth of coliforms. At relatively high concentrations (> 1 mg C/l) however, maximum growth rates of both coliforms and indigenous bacteria do not differ much. At such concentration levels coliforms would be able to compete more effectively with the indigenous bacteria provided that the water temperature is high enough. High concentrations of easily available substrates (carbohydrates) are not expected in drinking water. Therefore, it is much more likely that coliforms multiply in drinking water distribution systems when the concentration of favourite compounds suddenly exceeds the S_{min} value. The time period between this sudden increase and the moment that the indigenous bacteria have eliminated the substrates may be sufficient for coliforms to double their numbers several times. These numbers may then exceed the critical value of 1 CFU/100 ml. Subsequently, coliforms surviving in the drinking water distribution systems e.g. in the biofilm may cause non compliance for a relatively long period of time. Coliform regrowth also depends on water temperature. LeChevallier *et al.* (1991) found that most coliform occurrences in a distribution system were associated with water temperatures greater than 15°C.

The question remains which conditions cause (local) increases of the concentration of substrates available for coliforms. Treatment failure and/or a change in raw water quality water might give rise to an increase of the AOC concentration. LeChevallier *et al.* (1991) observed coliform problems when AOC concentrations in water leaving the treatment facility were greater than 50 μg of C/l. Furthermore, he found that *E. coli* was unable to multiply at an AOC concentration of 50 μg C/l (LeChevallier *et al.* 1987). Other investigators, aiming at defining an AOC-criterion for drinking water conducted growth measurements with coliform cultures in relation with AOC determinations. Rice *et al.* (1991) observed a statistically significant correlation between AOC concentration and coliform growth, and concluded that the coliform growth response increased with increasing AOC concentration. However, no AOC criterion could be derived from these data.

The introduction of biodegradable compounds (e.g. lubricants) or processes in biofilms and sediments may increase the substrate concentrations in certain areas of the distribution systems, even at low AOC concentrations. Sediments apparently are potential substrates for coliform growth and it is possible that conditions such as rise in water temperature, or local anaerobiosis on the pipe wall promote the release of growth substrates in sediment rich environments. It is not yet possible to describe these complex processes in detail. However, as already stated by Baylis (1930) more than 60 years ago, sufficient evidence is available to conclude that biofilm formation and sediment accumulation in drinking water distribution systems should be limited.

Recently, the biofilm formation characteristics of drinking water in the Netherlands have been determined using the biofilm monitor (Van der Kooij *et al.* 1995). Biofilm formation rates below 1 pg adenosinetriphosphate (ATP)/cm^2.day are typical for biologically-stable drinking water e.g. slow sand filtrate and aerobic ground water. At levels above 10 µg ATP/cm^2.day, regrowth of aeromonads was observed in drinking water during distributions with numbers exceeding the guideline value of 200 CFU/100 ml (Van der Kooij *et al* 1995). Investigations on the relationship between the biofilm-formation characteristics of drinking water and the frequency of coliform occurrences may lead to a criterion for biofilm formation, below which multiplication of coliforms in distribution systems is limited.

Acknowledgements

This report is written as part of the VEWIN-Research Program conducted by Kiwa NV. The author is much indebted to Peter Hiemstra and Geo Bakker (Water Supply Company of Overijssel) for providing data on coliform occurrences and coliform types in ground water supplies. The author acknowledges the assistance of Wim Hijnen and Corrie Gorte in conducting the measurements of growth and survival of the Klebsiella isolate.

References

Adler J, Hazelbauer GL, and Dahl MM. Chemotaxis towards sugars in *Escherichia coli*. J. Appl. Bacteriol. 1973;115:824-847.

Baylis JR. Bacterial aftergrowths in water distribution systems. Am. J. Publ. Hlth. 1930;20:485-489.

Baylis JR. Bacterial aftergrowths in water distribution systems. Water Works and Sewerage 1930; 335-338.

Burman NP and Colbourne JS. Techniques for the assessment of growth of micro-organisms on plumbing materials used in contact with potable water supplies. J. Appl. Bacteriol. 1977;43:137-144.

Camper AK, McFeters GA, Characklis WG and Jones WL. Growth kinetics of coliform bacteria under conditions relevant to drinking water distribution systems. Appl. Environ. Microbiol. 1991;57:2233-2239.

Dahl MK and Manson MD. Interspecific reconstitution of maltose transport and chemotaxis in *Escherichia coli* with maltose-binding protein from various enteric bacteria. J Bacteriol. 1985;164:1057-1063.

Drinking Water Inspectorate. Drinking water 1992. A report by the chief inspector, Drinking Water Inspectorate. Department of Environment, July 1993.

Flint KP. The long-term survival of *Escherichia coli* in river water. J. Appl. Bacteriol. 1987;63:261-270.

Geldreich EE, and Rice EW. Occurrence, significance and detection of *Klebsiella* in water systems. J. Am. Water Works Assoc. 1987;79:74-80.

Hedblom ML and Adler J. Chemotactic response of *Escherichia coli* to chemically synthesized amino acids. J. Appl. Bacteriol. 1983;155:1463-1466.

Herson DS, McGonigle B, Payer MA, and Baker KH. Attachment as a factor in the protection of *Enterobacter cloacae* from chlorination. Appl. Environ. Microbiol. 1987;53:1178-1180.

Herson DS, Marshall DR, Baker KH and Victoreen HT. Association of microorganisms with surfaces in distribution systems. J. Am. Water Works Assoc. 1991;83:103-106.

Hayashi S, and Li ECC. Capture of glycerol by cells of *Escherichia coli*. Biochemica et Biophysica Acta 1965;94:479-487.

Huck PM. Measurement of biodegradable organic matter and bacterial growth potential in drinking water. J. Am. Water Works. Assoc. 1990;82:78-86.

Joret JC et Levy Y. Méethode rapide d'évaluation du carbone eliminable des eaux par voie biologique. Trib Cebedeau 1986;39:3-9.

Kay WW and Kornberg HL. The uptake of C4-dicarboxylic acids by *Escherichia coli*. Eur. J. Biochem. 1974;18:274-281.

Krieg NR and Holt JG. Bergey's manual of systematic bacteriology, 1983; Vol. 1. Williams & Wilkins, Baltimore/London.

LeChevallier MW, Babcock TM and Lee RG. Examination and characterization of distribution system biofilms. Appl. Environ. Microbiol. 1987;53:2714-2724.

LeChevallier MW, Cawthon CD and Lee RG. Inactivation of biofilm bacteria. Appl. Environ. Microbiol. 1988;54:2492-2499.

LeChevallier MW. Coliform regrowth in drinking water: a review. J. Am Water Works Assoc. 1990;82:74-86.

LeChevallier MW, Schulz W and Lee RG. Bacterial nutrients in drinking water. Appl. Environ. Microbiol. 1991;57:857-862.

Lo TCY, Rayman MK, and Sanwal BD. Transport of succinate in *Escherichia coli*. J. Biol. Chem. 1972;247:6323-6331.

Martin RS, Gates WH, Tobin RS, Grantham D, Sumarah R, Wolfe P and Forestall P. J. Am. Water Works Assoc. 1982;74:(January):34-37.

Nobel PJ, te Welscher RAG, Hoogenboezem W. Medema GJ and Schellart JA.Bacteriën van de coligroep in drinkwater. Achtergrondinformatie en leidraad voor nader onderzoek. Kiwa rapport SWE 95.020. Nieuwegein 1995.

Piperno JR and Oxender DL. Amino acid transport systems in Escherichia coli K12. J. Biological Chemistry 1968;243:5914-5920.

Rice EW, Scarpino PV, Reasoner DJ, Logsdon GS, and Wild DK. Correlation of coliform growth response with other water quality parameters. Jour. Am. Water Works Assoc. 1991;83:98-102.

Ridgway HF and Olson BH. Chlorine resistance patterns of bacteria from two drinking water systems. Appl. Environ Microbiol. 1982;44:972-987.

Roberts DP, Sheets CJ and Hartung JS. Evidence for proliferation of *Enterobacter cloacae on* carbohydrates in cucumber and pea spermosphere. Can. J. Microbiol. 1992;38:1128-1134.

Seidler RJ, Morrow JE, and Bagley ST. Appl. Environ. Microbiol. 1977;33: 893-900.

Smith DB, Hess AF and Hubbs SA. Survey of distribution system coliforms in the United States. Proc. 1989 AWWA Water Quality Technology Conference; San Diego pp. 1103-1115.

Servais P, Billen G, Rego JV. Rate of bacterial mortality in aquatic environments. Appl. Environ. Microbiol. 1985;49:1448-1454.

Servais P, Anzil A and Ventresque C. Simple method for determination of biodegradable dissolved organic carbon in water. Appl. Environ. Microbiol. 1989;55:2732-2734.

Silhavy TJ, Ferenci T and Boos W. Sugar transport systems in *Escherichia coli*. *In* Bacterial Transport ed B.P. Rosen, 1987; p. 127-169. Microbiology Series Vol.4; Marcel Dekker, New York/Basel.

Talbot HW, Morrow JE and Seidler RJ. Control of coliform bacteria in finished drinking water stored in redwood tanks. J. Am. Water Works Assoc. 1979;71:(June)349-353.

Van der Kooij D, Visser A and Oranje JP. Multiplication of fluorescent pseudomonads at low substrate concentrations in tap water. Antonie van Leeuwenhoek, J. Microbiol. 1982; 48:229-243.

Van der Kooij D and Hijnen WAM. Regrowth of bacteria on assimilable organic carbon in drinking water. J. Francais d'Hydrology 1985;16: 201-218.

Van der Kooij D and Hijnen WAM. Multiplication of a *Klebsiella pneumoniae* strain in water at low concentrations of substrates. Wat. Sci. Tech. 1988;20:117-123.

Van der Kooij D and Hijnen, WAM. Determination of the concentration of maltose- and starch-like compounds in drinking water by growth measurements with a well-defined strain of a *Flavobacterium* species. Appl. Environ. Microbiol. 1985;49:765-771.

Van der Kooij D and Hijnen, WAM. Nutritional versatility and growth kinetics of an *Aeromonas hydrophila* strain isolated from drinking water. Appl. Environ. Microbiol. 1988;54:2842-2851.

Van der Kooij D. Assimilable organic carbon (AOC) in drinking water. *In* Drinking water microbiology; progress and recent developments (ed G.A. McFeters). pp. 57-87; Springer Verlag New York Inc, 1990.

Van der Kooij D. Assimilable organic carbon as an indicator of bacterial regrowth. J. Am. Water Works Assoc. 1992;84:(February): 57-65.

Van der Kooij D. Biofilm formation on surfaces of glass and teflon exposed to treated water. Wat. Res. 1995; 29: 1655-1662.

Van der Kooij D, Vrouwenvelder JS, Veenendaal HR and Van Raalte-Drewes MJC. Multiplication of *Aeromonas* in ground-water supplies in relation with the biofilm formation charatcteristics of drinking water. Proc.1994 Water quality Technology Conference. American Water Works Association Denver 1995.

Versteegh JFM, Van Gaalen FW and AJH van Breemen 1995. De kwaliteit van het drinkwater in Nederland in 1993. Ministerie van Volkshuisvesting, Ruimtelijke Ordening en Milieubeheer.

Von Meyenburg K. Transport-limited growth rates in a mutant of *Escherichia coli*. J. Bacteriol. 1971;107:878-888.

Wierenga JT. Recovery of coliforms in the presence of a free chlorine residual. J. Am. Water Works Assoc. 1985;78:83-88.

Chapter 23
Drinking Water Quality Regulation - A European Perspective

W M Waite

Principal Inspector, Drinking Water Inspectorate

1 INTRODUCTION

The purpose of regulating drinking water quality should be to ensure that the consumer receives water which is safe to use for normal domestic purposes over a lifetime and is aesthetically acceptable. Regulations should have a sound scientific basis and strike an appropriate balance between risk and cost. With these points in mind, this paper looks at drinking water quality regulation in England and Wales in the context of the United Kingdom's membership of the European Community. Because the subject of the conference is microbiology, the paper concentrates on regulation of the microbiological quality of drinking water and does not generally address physico-chemical quality aspects. The paper does not cover the regulation of bottled waters, natural mineral waters or medicinal waters.

2 THE BASIS OF MICROBIOLOGICAL STANDARDS

The demonstration by Robert Koch (Koch, 1877) that bacteria could be the cause of at least some diseases, and his development of solid media for the isolation of pure cultures of bacteria (Koch, 1881) opened the door some 110 years ago to bacteriology as a practical science. Frankland and others began to make sanitary assessments of waters using gelatine plates to count bacteria in waters, but it quickly became apparent that simple estimates of general bacterial content did not correlate well with the likelihood of a water source being associated with human infection. von Frisch has been cited as proposing as early as 1880 that faecal pollution of water could be revealed by the presence of organisms found in faeces. Escherich (Escherich, 1885) described several organisms in the faeces of newborn babies including a motile rod shaped organism capable of clotting milk which he named Bacterium coli commune. The affix commune was not generally retained and Escherich's organism became generally known as Bact. coli. Other workers found that Bact. coli was universally present in human faeces and Schardinger (Schardinger, 1892) proposed that because Bact. coli was always present in faeces, and had not at that time been demonstrated other than from faeces, its presence in water could be taken as an indication of faecal pollution of the water and therefore the possible presence of other pathogenic organisms.

The basis of Schardinger's principle was that examination of water for the presence of an organism invariably and only present in faecal matter gives an indication of the potential presence of enteric pathogens. Examination of waters for Bact. coli as an indicator of the potential presence of enteric pathogens was widely adopted using gas production from lactose for its identification. It soon became apparent that Bact. coli as determined was not a single species but a group of similar organisms, not all of which were necessarily associated with faecal pollution. When taxonomy and methodology developed and bacteriologists became able to recognise *Escherichia coli* (*E. coli*), the species which most closely matches Schardinger's proposed indicator, by being invariably and exclusively associated with faeces, they had a long tradition of monitoring for Bact. coli. Instead of moving over to monitor for *E. coli* rather than Bact. coli, which in the light of the underlying rationale would have been the correct thing to do, water bacteriologists began to monitor for *E. coli* in addition to Bact coli, acknowledging the heterogeneity of Bact. coli by adopting the term "coliform bacteria".

Bonde (Bonde, 1966) has observed that *"hesitancy often results in the adoption of procedures 'for the sake of security' and such procedures may be difficult to get rid of later on when they have lost all importance. It seems that this has happened when the coliform index was adopted instead of the Escherichia coli or faecal coliform index"*. This author wholeheartedly concurs with that statement.

For pragmatic rather than scientific reasons the practice of monitoring for the so-called thermotolerant or faecal coliforms rather than the faecally-specific *E. coli* has gained widespread adoption.

With Schardinger's principle in mind, other bacterial species and groups have been put forward as being characteristic of faecal pollution, two of the earliest being faecal streptococci and *Clostridium perfringens* (formerly *Cl. welchii*).

The rationale behind all of the so-called indicator bacteria is that they are characteristically faecal in origin and therefore if detected they indicate the presence of faecal pollution and the possible presence of pathogens. Mossel (Mossel 1978) attributed to Ingram the introduction of the concept of the index organism, which he described as an organism whose presence points to the possible presence of pathogenic similar organisms, and the indicator organism which indicates the failure of Good Manufacturing Practice affecting the final product. Using Ingram's terminology, the original water bacteriologists were using Bact. coli as an index of faecal pollution of water sources, a role for which *E. coli* is particularly suited, but coliforms much less so because of the environmental origin of many species and their ability to replicate in water, soils, on vegetation etc.

Water bacteriologists uncritically extended the use of coliforms to the monitoring of treatment processes and the distribution of water supplies, without considering whether Schardinger's principle still applied in these circumstances. Waite (Waite 1985)(Waite 1991) has argued for a fundamental review of the basis of bacteriological monitoring of drinking water and proposed that *E. coli* rather than coliforms is the best index organism available for assessing raw water quality, total Enterobacteriaceae rather than coliforms are the better indicator of the effectiveness of treatment (but are inadequate for assessing removal of Cryptosporidia and Giardia, for which no suitable candidate indicator bacteria have been identified), and *E. coli* is the best indicator of post-treatment faecal contamination. If a measure of post-treatment biological growth is required, Waite has proposed that non-selective colony counts and total Enterobacteriaceae are more suitable indicators than either coliforms or *E. coli*.

3 THE EUROPEAN DIMENSION

Europe is first and foremost a continent. Many, but not all, European countries have become members of the European Community, an economic and political union of 15 autonomous member states which have agreed to be bound by common aims and practices. The Community adopts Directives which do not themselves have legal application in the laws of member states and have to be given effect by the passing of appropriate national laws. Failure to pass suitable laws can however result in the member state being held to account by the European Court.

All member states are required to implement Directive 80/778/EC on the Quality of Water Intended for Human Consumption (EEC 1980). Member States are however permitted to supplement the Directive's requirements and set tighter standards if they wish. Directive 80/778/EC has been implemented in England and Wales by the Water Supply (Water Quality) Regulations 1989 (HMSO 1989) and by equivalent but separate Regulations in Scotland (HMSO 1990), and Northern Ireland (HMSO 1994a).

3.1 Directive on the Quality of Water intended for Human Consumption (80/778/EC)

The Directive lists 62 parameters and for the majority of these it prescribes Guide levels (GL), which are essentially advisory, and Maximum Admissible Concentrations (MAC), which are mandatory. There are a number of parameters for which no values are prescribed and Member States have discretion as to whether to set national standards for these. The microbiological parameters have MACs which are given in Table E of the Directive and are shown in Table 23.1. The Directive also specifies minimum sampling frequencies which apply at the point at which the water is made available to the user, the sampling points to be determined by the competent national authorities. In the UK water is deemed to be made available to the user at the point at which it passes from pipes owned by the water supplier into pipes owned by the consumer, that is to say at the curtilage of the property. Because it is not practicable to sample at that point, regulatory bacteriological sampling has to take place at consumers' taps. At least 50% of samples must be taken from taps selected at random, the remainder can, if desired, be taken from specially selected consumers' taps designated for regular sampling.

3.2 Implementation in Member States

Because the author has not been able to obtain copies of the regulations implementing 80/778/EC for every member state, and those obtained have not generally been in English translation, it is not possible within this paper to give a comprehensive overview. In particular the author has been unable to assess for most member states their requirements for sampling frequencies and locations. The author is aware that in at least part of one member state sampling takes place only from special sampling taps provided alongside meters measuring water entering premises, and not from normal domestic taps. Many UK water scientists contend that the kitchen tap makes a significant contribution to the numbers of samples found to contain coliforms, and gives a misleading impression of actual drinking water quality.

Despite the limited information available and the author's inability to understand the language of many of the texts, a number of differences between member states have been identified.

Table 23.2 summarises the author's understanding of the bacteriological standards adopted by the various member states. It is apparent that while regulation within the EC follows the basic principles of the Directive, there is a fair amount of difference in detail between member states. In particular, it is noteworthy that only Germany and Finland have specifically legislated for the monitoring of *E. coli* rather than faecal coliforms. The UK Regulations (HMSO 1989)

Table 23.1 *Microbiological Standards in Directive 80/778/EC*

Parameters	Sample vol (in ml)	Guide Level (GL)	Maximum admissible concentration (MAC)	
			Membrane filter	Mult. tube (MPN)
Total coliforms[1]	100	-	0	MPN<1
Faecal coliforms	100	-	0	MPN<1
Faecal strep.	100	-	0	MPN<1
Sulphite- reducing Clostridia	20	-	-	MPN≤1

Water intended for human consumption should not contain pathogenic organisms.
If it is necessary to supplement the microbiological analysis intended for human
consumption, the samples should be examined not only for the bacteria in Table
E but also for pathogens including:
- salmonella
- pathogenic staphylococci
- faecal bacteriophage
- entero-viruses
nor should such water contain:
- parasites
- algas
- other organisms such as animalcules
[1] Provided a sufficient number of samples is examined (95% consistent results)

Parameter		Sample size (in ml)	GL	MAC
Total bacteria counts for water supplied for human consumption	37°C (48hr)	1	10 [1,2]	
	22°C (72hr)	1	100 [1,2]	

[1] For disinfected water the corresponding values should be considerably lower
at the point at which it leaves the processing plant
[2] If, during successive sampling, any of these values is consistently exceeded a
check should be carried out

set a standard for faecal coliforms in line with the Directive and make no reference to *E. coli*.
However, in the reference document published by the Department of the Environment, which
is used as a Code of Good Practice by the Drinking Water Inspectorate (HMSO 1989 (b))
water analysts were expected to follow the recommendations in the report known colloquially
as Report 71 (HMSO 1982). This Report was produced jointly by the Department of the
Environment, the Department of Health and the Public Health Laboratory Service and while

Table 23.2 *Summary of microbiological standards for drinking water in member states*

State	Total coliforms (/100ml except where stated)	Faecal coliforms (/100ml except where stated)	Faecal streptococci (/100ml except where stated)	Sulphite reducing clostridia	Colony count 37°C/48 hours (except where stated)	Colony count 22°C/72 hours (except where stated)	Additional microbiological standards
Directive	0(95%)	0	0	≤1/20	GV 10	GV 100	
Austria (undisinfected)	0(95%)	0	0	≤1/20	10	100	Ps aeruginosa 0/100ml
(disinfected)	0/250	0/250	0/250	0/50	5	10	Ps aeruginosa 0/250ml
Belgium	0(95%)	0	0	0/20	used for regular surveillance	used for regular surveillance	
Denmark	0(95%)	0	0	0/50	20 (5 ex works)	200 at 21°C / 50(WTW, undisinfected) / 10 (WTW, disinfected)	
Eire	0 (MPN<1)	0 (MPN<1)	0 (MPN<1)	<1/20	No significant increase	No significant increase	
Finland	<1(95%)	<1 (E coli presumptive)	<1 (37°C, 48 hrs)	<1/20	GV 10	GV 100	
France	0(95%)	0	0	≤1/20	≤20 at 37°C (24hrs)	≤100 at 22°C (72hrs)	Pathogens including Salmonella 0/5 litre Enterovirus 0/10 litre Pathogenic staphylococci 0/100ml
Germany	0(95%)	0 (E coli)	0	no std	≤100, 36±1°C, 44±4hrs	≤100, 20±2°C, 44±4hrs ≤20 ex works after disinfection	
Greece	0(95%)	0	0	≤1/20	GV 10	GV 100	
Italy	0(95%)	0	0	0/100	GV 10 at 36°C	GV 100	
Luxemburg	0(95%)	0	0	0/20	GV 2 ex works 10 as supplied	GV 20 ex works 100 as supplied 20°C	
Netherlands	<1/300 ex works <1/100 in supply	<1/300 ex works <1/100 in supply	0	0/100	<10 in supply, annual geometric mean	<100 in supply, annual geometric mean	
Portugal	0(95%)	0	0	≤1/20	GV 10	GV 100	
Spain	0(95%)	0	0	≤1/20	GV 10	GV 100	
Sweden	0(95%)	0	0	≤1/20	GV 10	GV 100	
United Kingdom	0(95%) 0 ex works	0	0	≤1/20	No significant increase	No significant increase	

having no legal standing it has generally been accepted as the embodiment of best practice in the UK. The Report has subsequently been revised and republished (HMSO 1994b). The underlying rationale and philosophy of the 1982 edition of Report 71 was the enumeration of coliforms and *E. coli* and only passing reference was made to the concept of faecal coliforms. The practical consequence of this was that although the regulatory standard parameter was faecal coliforms, laboratories were expected to carry out analysis for *E. coli* as recommended by Report 71. The 1994 revision of Report 71 still focuses on the enumeration of *E. coli* and states that *E. coli* is a faecal coliform. It appears likely that UK practice will continue to be to monitor drinking waters for *E. coli*, as it has been for about the past 50 years.

It is also of note that the Netherlands have in general set more stringent standards than required by the Directive, as has Austria for disinfected supplies. In addition, France has set numerical standards for a number of the organisms listed in the footnote to Table E of the Directive, and Austria has set a standard for *Pseudomonas aeruginosa*.

3.3 The question of definitions

None of the microbiological parameters in the Directive are taxonomically defined and yet no definitions are given. There is an inherent, but ill-founded, assumption of universal agreement as to what constitutes a coliform, a faecal coliform, a faecal streptococcus etc. Waite (Waite 1985 p343) has contended that *the coliform group is more defined by the original techniques for its enumeration than the techniques for its enumeration based on a clear definition of the group* and that this has been a barrier to the development of new methods of analysis. For example by defining a coliform phenotypically as producing gas from lactose (Report 71 1982), the bacteriologist has been tied to the use of a cultural procedure involving the fermentation of lactose in a medium capable of demonstrating gas production. This precludes the development of alternative methods based on enzymic assays, gene probes, antigen detection etc. The adoption of a genotypic definition of coliforms (HMSO 1989b) (HMSO 1994b) has removed this restriction in the UK and given the coliform a scientifically defensible definition. The definition of coliforms now adopted is *members of genera or species within the Family Enterobacteriaceae, capable of growth at 37 °C, that possess β-galactosidase.*

There is an equally strong case to be made for replacing the faecal coliform, which is defined primarily by its ability to ferment lactose at 44°C, by $E. coli$, a taxonomically defined species which better satisfies Schardinger's principle. Similarly, the sulphite-reducing Clostridia parameter is less suitable than the organism *Clostridium perfringens* It is suggested that a similar genotypic rather than phenotypic definition for faecal streptococci, or perhaps enterococci, is also needed in order to put the current regulatory parameters on to a more scientifically robust footing.

3.4 Proposed revision of Directive 80/778/EC

The European Commission has determined that there is a need to revise Directive 80/778/EC in the light of scientific developments in the 20 years since its drafting. The long process of consultation and negotiation is just beginning, and the form of the final revision is by no means certain, but the draft proposals contain a number of fairly fundamental changes to the microbiological aspects of the current Directive. These are:

(a) Parameters should be divided into 3 groups, Microbiological, Chemical and Indicator. Microbiological and Chemical parameters are to be included because of their direct significance for the protection of human health and member states must set standard at least as strict as those in the Directive. Indicator parameters do not in themselves at the values proposed present a risk to human health but are included to provide a prompt indication of changes in water quality and of the possible need for action to protect public health. Although values are prescribed for these indicator parameters, compliance with those values would not be compulsory. The values would be trigger values beyond which it is necessary to confirm that water quality is adequate to protect human health. Only if it is inadequate is further action needed. Member states would be permitted to set less stringent values for indicator parameters appropriate to local circumstances. The coliform parameter is included in the indicator parameter group. This means that although standards would be set for coliforms and monitoring required, providing further investigations after detection of coliforms did not reveal the need for action to protect human health, the detection of coliforms would not in itself be a breach of the requirements of the Directive. If accepted this would permit due allowance to be made for the occurrence of coliforms when they are not derived from faecal pollution.

(b)   The standards should be complied with at at least one tap in the consumer's premises, but should not apply to in-house storage systems and hot water systems. This would clarify the currently vague Directive term "point at which it is made available to the user" and at the same time not make the state and supplier responsible for breaches of the standard due to the domestic plumbing which is the responsibility of the property owner.

(c)   The residual chlorine parameter should be replaced by a requirement to monitor to verify the efficiency of disinfection. This takes account of the use of alternative disinfectants and focuses on the main purpose behind the original adoption of the residual chlorine parameter. It does not however take account of the role of chlorine in the distribution system as a putative inhibitor of biofilms and indicator of ingress or aftergrowth.

(d)   The Directive should only set the minimum monitoring requirements and leave member states to set more stringent monitoring requirements as appropriate. The sampling frequencies proposed for *E. coli* are the same as in the current Directive for faecal coliforms, but without the proviso that where water has to be disinfected, the sampling frequency should be doubled. This could open the door to a significant reduction in drinking water monitoring.

(e)   Bacteriological sampling points should be taps within domestic premises. This will ensure greater consistency between member states monitoring practice, bringing them into line with current UK practice.

(f)   Replacement of the faecal coliforms parameter by *E. coli*. The benefits of this change have been discussed above.

(g)   The recommended conditions for colony counts would change from 48 hrs at 37°C and 72 hours at 22°C to 21 hours at 37°C and 72 hours at 21°C. There are no reasons given in support of this change.

(h)   Sampling frequencies should be based only on volume of water supplied.

(i)   There should be a check monitoring frequency set for a suite of key parameters, and a lower audit monitoring frequency for all other parameters. *E. coli* is in the check monitoring suite, whereas coliforms, faecal streptococci, sulphite-reducing Clostridia and colony counts are covered by audit monitoring only. It should be noted that the audit monitoring frequency applies not only to indicator parameters but also to some microbiological parameters (see (a) above).

4 COST BENEFIT CONSIDERATIONS OF MONITORING FOR COLIFORMS IN DRINKING WATER DISTRIBUTION SYSTEMS.

It is the author's belief that almost invariably when coliforms are isolated from drinking water supplies without the detection of *E. coli* or faecal coliforms, their presence is due to something other than the ingress of faecal pollution and is not indicative of a significant risk to the health of consumers. The setting of a 95% compliance requirement for the coliform standard, when no such qualification is placed on the faecal coliform standard, emphasises the lower importance attached to this parameter.

There is a real cost associated with the implementation of the coliform standard, and that cost should be justifiable in terms of the benefit to the public health.

In 1994, 1633 of the 423,690 samples taken from service reservoirs and consumers' taps in England and Wales yielded coliforms but not *E. coli*. The response to a coliform failure in these cases would usually involve the taking of a minimum of three additional samples and their examination for coliforms and *E. coli* and also faecal streptococci. On occasion the resamples may also yield coliforms and a more far reaching response including operational action such as systematic flushing and raising of chlorine concentrations may be required. In extreme cases where coliforms have been found to occur sporadically and are suspected to be associated with biofilm growth, companies may have had to enter into undertakings to investigate causes and take action to eliminate them. There have been instances of companies issuing advice to boil water on the strength of the isolation of coliforms but not *E. coli* from a number of concurrent samples. The issue of advice to boil water may cause concern among consumers, and expose them to inconvenience, expense and the risk of scalding, not to mention possible psychosomatic illness.

One water company has estimated the minimum cost of responding to a coliform detection at a consumers tap as £200 and at a service reservoir as £250. In many instances costs will far exceed those figures, particularly if a positive resample is obtained. At a conservative estimate, the cost to water companies of the 1633 coliform isolations in the absence of *E. coli* in 1994 was more than £400,000. The author is aware of incidents involving the detection of coliforms in the absence of *E. coli* which have resulted in potential legal action on behalf of consumers, and in some of these cases total costs before any case has come to court may easily exceed £100,000. These events also make demands on the regulator and thereby incur additional costs.

It is worth debating whether this expenditure represents the best value for money in relation to protection of the public health.

5 PROS AND CONS OF FORMAL REGULATIONS

Prior to the EC Directive 80/778/EC the regulatory requirements for drinking water in England and Wales were simply that it be wholesome, and there were no legally binding requirements regarding monitoring practice. Report 71 recommended sampling frequencies for microbiological monitoring. With the coming in to force of the Directive the UK initially gave the Directive effect by simply stating that compliance with the Directive was a pre-requisite of wholesomeness (DOE 1982), and only in 1989 were specific regulations for drinking water quality promulgated (HMSO 1989). At that time the Drinking Water Inspectorate was set up and the current formal systems for publishing results of drinking water monitoring were established.

These Regulations have contributed immensely to the control and improvement of drinking water quality and ensured that compliance with standards is properly monitored and enforced in a publicly accountable manner.

In the past, water bacteriologists would concentrate their monitoring in known or suspected problem areas, and their primary concern was to protect the consumer by detecting and eliminating quality deficiencies. By the same token, there was an underlying drive to develop newer methods which would give faster results and be more sensitive. With the advent of the Regulations and public accountability the emphasis has shifted to satisfying the regulatory requirements as first priority. When the new definition of coliforms was adopted, and in consequence the gas production criterion which had been criticised as being scientifically unjustifiable (Waite 1985) was deleted, there was opposition from a number of water companies. This was understandable because their concern was that the new definition of coliforms included anaerogenic organisms which hitherto had been considered as presumptive coliforms which failed to confirm. Waite has argued that anaerogenic coliforms are of no more nor less sanitary significance than aerogenic ones, and therefore the new definition has made the monitoring regime more robust. The change however raised the spectre of more samples not complying with the standard, and an apparent deterioration in water quality being made public. Similar concerns may well apply between member states with the introduction of the Directive 91/692/EEC Standardising and Rationalising Reports on the Implementation of certain Directives relating to the Environment (EEC 1991) since many different methods for coliform enumeration are currently in use and are acceptable under Directive 80/778/EC. Member states could well be reluctant to change their methods for those of another member state even if there was good evidence that the alternative method was more sensitive, because that could make the State's water quality appear to deteriorate.

This author believes that both regulator and regulated must ensure that initiatives to increase the sensitivity of methods or otherwise increase the protection of the health of consumers are not jeopardised by the potential implications for regulatory compliance. The purpose of bacteriological monitoring of drinking water is the protection of public health, and regulations set standards and provide the framework which ensures that appropriate monitoring is carried out. Compliance with regulations is a means by which the end of protection of public health is achieved, but should not be allowed to become an end in itself and the sole aim of monitoring.

Acknowledgements

The author wishes to thank Michael Rouse, Chief Drinking Water Inspector, for permission to publish this paper and wishes to make it clear that the ideas expressed in it are his own and must not in any way be attributed to the Inspectorate. While questioning some of the basis of the current regulatory framework, the author remains committed in his role as a Drinking Water Inspector to ensuring that water companies comply with all of their regulatory duties within that framework.

References

Bonde G.J.(1966) Bacteriological methods for estimation of water pollution. Health Lab. Sci. 1966;3:124

EEC (1980) Directive relating to the quality of water intended for human consumption. 80/778/EC. Official Journal L229

EEC (1991) Directive 91/692/EEC Standardising and Rationalising Reports on the Implementation of certain Directives relating to the Environment 90/692/EC

Escherich T.(1885) Die Darmbakterien des Neugeborenen und Säuglings. Fortschr. Med. 1885; 3(16):515- and 547-

HMSO (1982) Reports on Public Health and Medical Subjects No.71. The Bacteriological Examination of Drinking Water Supplies 1982

HMSO (1989a) Water Supply (Water Quality) Regulations 1989. Statutory Instrument 1989 No 1147

HMSO (1989b) Guidance on Safeguarding the Quality of Public Water Supplies

HMSO (1990) Water Supply (Water Quality)(Scotland) Regulations 1990. Statutory Instrument 1990 No 119 (S 11)

HMSO (1994a) The Water Quality Regulations (Northern Ireland) 1994. Statutory Rules of Northern Ireland No 221

HMSO (1994b)Reports on Public Health and Medical Subjects No.71. The Microbiology of Water 1994 Part 1-Drinking Water

Koch R. (1877) Untersuchungen über Bakterien V. Die Aetiologie der Milzbrand-Krankheit, begründet auf die Entwicklungsgeschichte des Bacillus Anthracis. Beitr. Biol. Pflantzen 1877; 2(2):277-310 Translation in Brock T, Milestones in Microbiology; Amer. Soc. Microbiol: 1975

Koch R. (1881) Zur Untersuchung von pathogenen Organismen. Mittheil. Kais. Gesund. 1881;1:1-48 Translation in Brock T, Milestones in Microbiology; Amer. Soc. Microbiol: 1975

Mossel D.A.A. (1978) Index and indicator organisms; a current assessment of their usefulness and significance. Food Technol. Aust. 1978;30:212-219

Waite W.M. (1985) A critical appraisal of the coliform test. J. Inst. Wat. Eng. and Sci. 1985;39:341-57

Waite W.M. (1991) Drinking Water Standards - a personal perspective. In:Proceedings of the UK Symposium on Health-related Water Microbiology. 1991 (Editors Morris R. et al) 52-65. London, International Association for Water Pollution Research and Control

Chapter 24
Reinventing Microbial Regulations for Safer Water Supplies

Edwin E. Geldreich

Senior Research Microbiologist, National Risk Management Research Laboratory, Cincinnati, Ohio, 45268

1 INTRODUCTION

Providing safe water supply through regulations is a difficult task. Utilities complain of excessive oversight on their professional expertise while the increasing public awareness of health issues causes a loss of confidence with public water supply whenever there is a non-compliance problem with water quality. Recently the EPA Administrator (Carol Browner) reported that one in five public water supplies in the United States does not meet the existing regulations. If this is the case, can regulations be written that protect all public water supplies? There are three major problem areas that make the task difficult: patterns of waterborne outbreak occurrences, weaknesses of small system compliance and the microbial risk issues.

2 MICROBIAL RISKS

Enhanced activity in detecting and reporting waterborne outbreaks has demonstrated that most of these occurrences are to be found in the small water systems, although large utilities are not immune to outbreaks caused by marginal treatment practices. Some startling facts emerge from the reports of waterborne outbreaks in the United States (Table 24.1) over the past 54 years. By far the largest outbreak involved *Cryptosporidium* with over 370,000 illnesses among a population of 800,000 in Milwaukee Wisconsin. *Giardia* has been involved in 84 outbreaks most of which occurred in supplies using remote watersheds and disinfection as the sole treatment. Among the bacterial pathogens, *Salmonella* species were responsible for over 25,000 cases in 38 outbreaks, and *E. coli* 0157:H7 caused 600 of 1,100 people to become ill in a small community.

Microbial risk assessment is necessary for the development of better regulations that are sensitive to health issues at a price that is cost effective through treatment. The problem is more complicated than risk assessment procedures proposed for chemicals because occurrence of a pathogenic agent in water will always be variable and the end points may range from no disease for a few, to infection for some consumers and possible death for sensitive segments in the community. In other words, the health status of the consumer plays an important part in the possible outcome of exposure. What is most disturbing is that the causative agent for approximately half of all U.S. outbreaks has never been identified, partly because of the logistics necessary to initiate a prompt epidemiological investigation, but also because there may be agents that are undetected by current laboratory procedures. As a consequence, risk

Table 24.1 *Waterborne Outbreaks of Disease in the United States of America * (Period 1940 - 1994)*

Etiologic agents	Outbreaks	Cases	Deaths
Bacterial			
Shigella	52	7,462	6
Salmonella	38	25,286	8
Campylobacter	5	4,773	0
Toxigenic *E. Coli*	6	1,431	8
Cholera	1	17	0
Vibrio	1	17	0
Yersinia	1	16	0
Viral			
Hepatitis A	51	1,626	1
Norwalk	16	3,973	0
Rotavirus	1	1,761	0
Protozoan			
Giardia	84	22,897	0
Entamoeba	3	39	2
Cryptosporidium	3	418,000	2
Chemical			
Inorganic (metals, nitrate)	29	891	0
Organic (pesticides, herbicides)	21	2,725	7
Unidentified agents	266	86,740	0
Total	**578**	**613, 417**	**34**

* Sources:
  Lippy and Waltrip, 1984
  Adapted from Craun, 1985
  Hayes, et al., 1989
  Geldreich, et al., 1992, 1994

assessments based on one infected individual in a population of 10,000, taking into consideration the causative agent, cost-effective treatment required and appropriate monitoring strategy, will be difficult to establish. All of these investigations assume that only one pathogenic agent is involved in every outbreak. This is a misconception. Sewage often has a multiple pathogen content, so it is possible that more than one pathogen may be involved in a waterborne outbreak that had its origins in a sewage contamination. Another complicating factor is pathogen testing sensitivity; some procedures have recovery efficiencies of 30 percent or less, providing no clue as to viability and infectious state.

Opportunistic pathogens introduce another element to the microbial risk issues. A study of community health effects related to water supply in the Quebec area suggests that there can be a measurable contribution to waterborne disease from opportunistic pathogens growing in distribution systems (Payment *et al.*,1991). Those at risk are sensitive segments of the population: young children, senior citizens, victims of AIDS, patients receiving chemotherapy, and individuals with kidney disorders. This fact introduces a management dilemma on who should be protected from risks — the average healthy group of people in the community or those individuals in the higher risk category. Current policy is to set the regulations for minimizing the health risks to the average healthy majority in the population, but to develop health advisory information for utilities and the medical community to pass on to those individuals in the higher risk category. First, however, the health concerns must be identified, then a limit set, based on attainability through cost effective treatment. Little is known at this time about the specific opportunistic organisms of concern, their virulence and most effective way to control them, whether through assimilable organic carbon (AOC) reduction, cyclic flushing schemes or application of alternative disinfectants.

3 MONITORING CONCERNS

Fundamental flaws in public health protection appear to be ineffective monitoring, inadequate response to coliform occurrences, insufficient treatment barriers and inefficient protection from contaminate intrusions in distribution systems (Table 24.2). Often these situations are triggered by a combination of factors related to surface water that is subject to wide quality fluctuation caused by rainfall events over the watershed. These storms flush a variety of fecal contaminants — including pathogens from wildlife, domestic animals and man — into the water course. Groundwater is not immune to pollution either. There is growing evidence that groundwater quality protection is not always assured through soil barriers. Reports of some waterborne outbreaks and unsatisfactory coliform results in groundwater have led to the disinfection of many well waters and springs used for community water supply (Lippy and Waltrip, 1984; Craun, 1986).

Monitoring and appropriate follow-up remedial actions appear to be inadequate or neglected in small systems. These are the utilities that have no laboratory of their own and depend on the state monitoring service to provide water quality data which is limited to only a few samples per month, based on population served. Often, several weeks may pass before the small utility receives reports of water quality episodes that occurred in the recent past. While a historical record of past performance is useful, the lateness of the information provides no opportunity to make midstream corrections that might stop a contaminate passage to the consumer's tap.

Comprehensive monitoring of source water, treatment processes and water supply in distribution for all known pathogens is impossible for three reasons: (1) coliforms are not good estimators of some pathogen occurrences because of differences in source water survival rates; (2) sensitivities to various treatment processes are not uniform and (3); the persistence potential for various pathogens in distribution system biofilms is unknown. There are several alternatives: monitor for all known pathogens; select the most resistant pathogen for testing the performance of treatment-train configurations; or search for new indicator candidates such as aerobic sporeformers, acid-fast bacteria, yeast, or coliphage.

Table 24.2 *Water System Deficiencies Causing Disease Outbreaks, 1920-1983* *

Public Water System Deficiencies	Number of Systems	Percent of Deficiencies	Illness Cases
Contaminated, untreated groundwater	661	43.2 %	82,528
Inadequate or interrupted treatment	333	21.8 %	224,973
Distribution network problems	233	15.2 %	83,577
Contaminated, untreated surface water	158	10.3 %	12,709
Miscellaneous causes or insufficient evidence	146	9.5 %	11,542
Totals	**1,531**	**100.0 %**	**415,329**

* Data adapted from Craun (1985)

Sampling frequency formulae based on population served are the accepted practice in many countries (Environmental Protection Agency, 1976; World Health Organization, 1984; Department of Health and Social Security, 1969; Council of the European Communities, 1975; Ministry of National Health and Welfare, 1977). The basis for this approach is the recognition that as the population increases, so will the size and complexity of the system and the corresponding potential for distribution network contamination by cross connections and back siphonage (World Health Organization, 1970).

Sampling frequency should be adequate to characterize the water as continuously safe to drink. In reality, microbiological monitoring can test only a minute faction of all available water supply during some designated instance of time over a thirty day period. This is indeed a difficult quality control task, considering the number of variables that can occur. Several factors must be taken into consideration if frequency of monitoring is to be optimized (Hoskins,1941; Technical Subcommittee, 1943; World Health Organization, 1971; Safe Drinking Water Committee, 1977;Berger and Argaman, 1983;Donner, 1987;and Committee on the Challenges of Modern Society NATO, 1984). For one, the integrity of the distribution system is often an important issue with old pipe networks. Systems that have been in service for over 75 years have acquired a variety of pipe materials, numerous potentials for cross connections, and opportunities for flow reversals as the system is expanded to meet a growing population. As a consequence, every water distribution system is unique in its strengths and weaknesses and therefore requires careful customizing of the monitoring frequency. In reality, many utilities collect more samples per month than are required by law (AWWA Committee on Bacteriological Sampling Frequency, 1985).

While the sampling frequency established has, in general, proven to be adequate for routine monitoring of water supplies serving populations over 10,000, the collection of fewer than 10 samples per month in smaller water systems leaves many hours of uncertain water quality among those utilities with the poorest record of compliance and the most waterborne outbreak occurrences. In these cases, it should be mandatory that at least one sample be taken from the first customer location each week. Collecting another sample each week near the end of a pipe line on the distribution system would serve as a check on static water quality in the small system pipe network. The habit of collecting a few samples each month from the same favored customer sites in the center of the distribution system provides little assurance that water quality is acceptable throughout the community (Environmental Protection Agency, 1989). While the pipe network of a small water system is less complicated, because of fewer service connections, there is always a danger that water quality may deteriorate from low water pressure and cross-connections.

Having established some specified number of sample locations per month, the next decision is when should these samples be collected during the month? It is logical to assume that the monitoring program should be spread out over the entire thirty day period because of the desire to continually check on the quality of potable water produced? In general, this is the approach most utilities use unless they are small water systems. Since small water systems are required by regulation to submit five or fewer samples per month, these sample collections are most frequently clustered on one day, with no other monitoring for bacteriological quality during the remaining days of the month. This minimal monitoring may be acceptable if a protected ground water supply is used, but could be less desirable if surface water is the source because of the greater variation in water quality.

The one aspect that is often difficult to resolve relates to appropriate time for sample collection. Theoretically, sample collection should be distributed uniformly over the month and timed to coincide with filter backwashing schedules in treatment or with distribution system maintenance including major repairs. This ideal approach is only occasionally achieved, more through coincidence than by intent. Small systems cluster their sample collections on one day early in the week (Table 24.3) for shipment to a distant laboratory. Larger utilities (serving over 2,500 people) do make an effort to collect samples throughout the month. These sample collections are most often made on Mondays and Tuesdays with a minimal number collected at the end of the week or on Sunday. This practice is a reflection of concern for shipment by various modes of parcel service (mail, bus, other delivery services) to minimize transit time over weekends and holidays. For those utilities that have their own laboratory, sample collections can be done more frequently during the week and at different times during the day. More remarkable is the practice of some utilities to collect samples in the evening or over-night (Table 24.4) for early morning shipment or processing at the start of laboratory hours. The more common procedure is to schedule sample collections in the morning for afternoon processing of shipment to a distant laboratory. The net effect of these patterns is that the samples are most likely to be collected at set times that do not vary and are not necessarily correlated with times when sand filters are returned to service, other adjustments are made in treatment practice, or related to pollution spills into the area of raw water intakes.

Table 24.3 *Distribution of Sample Collection During Week**

Population Served (Thousands)	Utilities Reporting That Samples Are Always Or Frequently Collected On Indicated Day - Percent						
	Sun	Mon	Tues	Wed	Thurs	Fri	Sat
<1	0	56.6	54.0	31.8	13.3	8.8	0
1 - 2.5	2.1	66.1	51.0	23.2	18.5	7.3	1.7
2.5 - 4.9	2.0	66.0	61.6	37.9	24.6	13.8	2.4
4.9 - 8.5	2.2	62.4	62.4	42.9	26.1	13.7	1.3
8.5 - 12.9	2.8	61.8	64.1	48.6	27.6	11.6	1.7
12.9 - 18.1	4.8	61.4	61.4	48.8	31.3	16.3	4.2
18.1 - 24	2.3	56.8	63.6	42.0	29.5	21.6	2.3
24 - 50	5.8	62.4	63.4	58.5	38.0	24.8	5.8
>50	14.7	83.5	84.1	76.0	64.8	48.2	14.8

* Data from AWWA Committee Report (2)

Table 24.4 *Distribution of Sample Collection During Day **

Population Served (Thousands)	Utilities Reporting that Samples are Always or Frequently Collected at Indicated Time - Percent			
	Morning	Afternoon	Evening	Night
<1	75.2	47.8	9.7	0
1-2.5	74.2	32.6	0	0
2.5-4.9	76.4	48.3	2.9	0.5
4.9-8.5	74.3	42.9	2.2	0
8.5-12.9	82.3	44.2	1.1	1.1
12.9-18.1	88.6	40.4	1.8	1.2
18.1-24	88	40.9	1.0	0
24-50	92.7	40.9	3.4	1.4
>50	98.2	65.8	4.9	3.9

* Data from AWWA Committee Report (2)

4 LABORATORY CREDIBILITY

Nothing could be more critical in the characterization of microbial quality of water supply than laboratory credibility. Critical decisions must often be made on evidence supplied from laboratory reports. Such data is often pivotal to a proper action response, to compliance with governmental regulations and, perhaps, during legal recourse following extended boil-water orders, illness or death resulting from a waterborne outbreak.

It must be remembered that the current program of laboratory certification in the United States focuses only on a narrow range of concerns with coliform occurrences and, to a limited extent, on the status of heterotrophic bacterial densities as a   verification of disinfection impact on water quality in distribution. Pathogen detection and speciation of these organisms and those of coliforms in biofilm are not recognized as part of the laboratory services required in routine monitoring. While this laboratory certification program is limited in coverage, the opportunity to have a scheduled review by State or Federal officials, every one-to-three years does have a positive influence on sustaining good laboratory practices for a variety of testing services. Experience has shown (Geldreich, 1967) that there is a continuing need for laboratory certification services both at the state and municipal level to keep the number of deviations to an absolute minimum. The optimum frequency of laboratory certifications at the state level appears to be 3 years. Visits to these laboratories at more frequent intervals yield little value to either the staff or the certification program; however, longer intervals result in an increased number of deviations. On-site evaluations of local laboratories should be made on a yearly basis because these facilities often have a more frequent turnover of technicians, and more equipment problems and space limitations.

Critical to the evaluation of a laboratory for certification is defining a group of fundamental requirements absolutely essential for data development. These criteria (Laboratory Certification Program Revision Committee, 1991) include laboratory facilities, equipment and supplies, basic laboratory practices (analytical methods, quality assurance/quality control), sample collection and handling, data handling and response to laboratory results. Some of these critical items are also among the most frequently occurring deviations found in all laboratories (Table 24.5).

Another crucial element to an effective monitoring program is the laboratory staff. They must be knowledgeable in the performance of a variety of microbiological testing procedures. Since drinking water microbiology has been rapidly expanding in the last decade, all laboratory staff members need to become involved in a continuing education program that focuses on advances in the microbiological analysis of drinking water. Of particular importance are those workshops dedicated to laboratory certification, pathogen detection and new advances in drinking water microbiology. It is important to note that sustained skills in drinking water microbiology can only be preserved by frequent analyses per month, not by performing a few tests per year or by citing a certification of past accomplishments.

Without proper quality control procedures, the reliability of the data generated is in question. Quality of commercial products used in the laboratory and stability of instrumentation and equipment must never be taken for granted. Even the most respected manufacturers may on occasion release a product that does not perform properly. Instrumentation reliability has improved greatly with the advent of solid-state electronics. However, there may be unpredicted failures in components due to voltage spikes, aging of sensing probes and deterioration in parameter specifications for solid-state devices that can cause drift in measurement accuracies. As a safeguard against these possibilities, every laboratory must have a quality assurance program that includes adequate quality control tests on analytical procedures, commercial supplies, instrumentation and equipment.

Table 24.5 *Most Commonly Occurring Basic Deviations from Standard Methods in Water Laboratories* *

Items	Percent Deviations	
	60 State & Regional Laboratories	70 Municipal Laboratories
Sample bottles (size, air space)	*56.7***	*30.1***
Distilled water suitability	*51.7***	*25.7***
Sample transit time	*45.0***	8.6
Buffered dilution water	*31.7***	*31.4***
Incubator (temp., record, Maintenance)	*28.3***	*65.7***
Clean glassware	18.3	*25.7***
Media pH (pH equipment, pH record)	15.0	*70.0***
Sample dechlorination	11.7	*42.9***
Autoclave (procedure, Maintenance)	11.7	*50.0***
Media sterilization	8.3	20.0
Media storage	1.7	*28.6***
Sampling frequency	1.7	12.5

* **Data revised from Geldreich**
** **Percents in *italics* donate most frequent deviations**

5 BIOFILM SIGNIFICANCE

Historically, the total coliform group was selected for assessing the microbial quality of drinking water because these bacteria are consistently present in great numbers and can be easily detected in highly diluted contaminated water. In recent years, emphasis on interpreting coliform occurrences has shifted from its possible sanitary significance to using coliforms as a measure of ground water protection and treatment effectiveness. Proper   processing of raw water should result in a six log reduction of coliforms bacteria.

Distribution network biofilm has introduced another twist to the coliform regulation. In recent years there has come a growing awareness that some environmental coliform strains can colonize the distribution system. Once heterotrophic organisms become established in the pipe environment, they will eventually form a biofilm community that may provide a haven for some coliform bacteria (*Enterobacter, Klebsiella, Citrobacter*) passing through the pipe network. Given an adequate nutrient base, in a protected pipe sediment or tubercle and a water temperature over 15°C, growth is accelerated. In time, biofilm fragments are torn away by the shearing force of water passage, flow reversals, and changes in the structure of pipe sediment due to water pH and coliform-positive samples become a reality. To what extent these biofilm occurrences become a major water quality problem will often be dictated by the maintenance program on the distribution system. Corrosion control, effective flushing programs, elimination of static water zones and maintenance of a disinfectant residual in 95 percent of the pipe network are important in reducing microbial colonization of the distribution network and seasonal threats of coliform biofilm releases.

While these coliform occurrences may not be of public health concern, they should not be ignored because the contamination suggests: a) existence of a habitat that could be used by pathogens; b) possible leaks in the treatment barrier or distribution system; and c) the presence of excessive AOC accumulation that interferes with maintaining a disinfectant residual throughout the pipe network.

In several cities (Muncie, IN; Springfield, IL; New Haven CT), where the frequency of coliform occurrences resulted in a non-compliance period, careful surveillance by local medical clinics and hospitals was made on all new patients to determine if there was any indication of a waterborne disease outbreak. No increase in the incidence of intestinal disease was noted. Similarly, during the period from April to September 1984, a review of nosocomial infections among intensive care patients of a large hospital in New Haven, CT, revealed no increase in *K. pneumoniae* infections from a gentamicin sensitive strain similar to the one isolated from the water supply (Ludwig *et al.*, 1985). More attempts need to be made to "fingerprint" biofilm isolates and compare these species patterns with ongoing clinical strains of opportunistic coliforms in hospitals using the same water supply during a coliform biofilm event.

There is always a concern that biofilm occurrences could be hiding a fecal contaminating event from inadequate treatment or distribution system contamination. For this reason, any utility experiencing coliform biofilm in the system should intensify the monitoring program and search for evidence of fecal coliforms or *E. coli* among the positive samples. If detected, verified occurrence of these coliforms should call for a boil-water order until repeat samples prove their disappearance. Any occurrence of fecal coliform bacteria or *E. coli* during a biofilm episode should not be brushed aside as an aberration merely because these organisms are not able to permanently colonize biofilms.

6 COMPLIANCE STRATEGIES

There are several major issues that must be considered in formulating regulations: attainability in the field, cost effectiveness, monitoring capability, enforcement potential (including public notification) and defining an acceptable consumer risk level. Originally, regulations were based on the study of data from a variety of water systems in the early 1900's. This data suggested that a utility supplying drinking water with an average coliform density of less than one organism in 100 ml of sample per month appeared to provide a safe drinking water. For water treatment plants using surface water, this goal could be achieved when the raw source water was treated by appropriate engineering processes or by other utilities who chose to use a protected ground water source. Several conditions were specified including: the sample size must equate to a glass of water (100 ml) and the number of samples to be examined per month from the distribution system should based on the size of the population served.

This historical monitoring approach focused on a demonstration of safe water quality in distribution to the consumer's tap, with lesser emphasis on monitoring treatment barrier effectiveness. Now the strategy is expanding to include sampling finished water entering the distribution system and sites at the first customer location. Inclusion of these permanent sampling sites illustrates the growing concern about diminished treatment barriers as efforts are made to minimize disinfection application in an effort to reduce by-product formation.

7 PUBLIC NOTIFICATION

Coliform occurrences in the distribution system should always be interpreted as evidence of a contaminating event (Geldreich and Rice, 1987). What needs to be determined first is whether such an occurrence represents fecal contamination, stressed coliform passage (McFeters *et al.*, 1986), or a release of coliforms that have colonized the distribution system. Any evidence of fecal coliform or *E. coli* in the repeat samples calls for an immediate public notification. If the fecal contamination is confined to one zone, then public notification can be restricted to only those customers in the affected area. When fecal contamination is system-wide, there must be an immediate alert to all customers in the community to boil water for drinking purposes until further notice. During this period, all resources must be utilized to find the cause and apply appropriate measures to restore water quality.

Public trust in the community water supply can be shaken by news reports on waterborne health hazards or by repeated public notifications of non-compliance with federal drinking water regulations. In recent years, U.S. regulations have placed greater emphasis on the public's right to know about risks associated with drinking water quality. While most people understand that life is not risk free, they do not normally associate their drinking water with risk to their well-being, nor do they understand much about water treatment and delivery to the residential tap. They are, therefore, confused and disturbed by sudden boil-water notices and by inaccurate statements on laboratory reports and risk assessments. As a consequence of these fears there is a small but growing segment of the public that is convinced that many health risks in community water supplies are non-reversible so that the only option available is bottled water or point-of-use water supply treatment in the home. To counter this growing concern, water utility professionals need to sharpen their skills at informing the public on water quality issues in a way that instills confidence, along with the recognition that a safe water supply may incur additional operating costs for infrastructure repair or additional treatment.

What are some of the realities about drinking water quality that the public should be aware of? Most important is a realization that the selection of criteria and standards applied to drinking water quality is not done in a capricious manor. The regulation process begins with early alerts to some new contaminant in water supply. These may come from several directions: industrial disasters involving new chemical exposures, accidental spills of exotic materials from truck or rail transport, waterborne outbreaks or discovery of new waterborne agents and awareness of pathogen adaptation to the aquatic ecosystem. Such information may be gathered from incidents in the United States or other nations, or through alerts from the World Health Organization and other international agencies participating in a global watch over environmental issues.

8 APPLIED RESEARCH FOCUS

Investigation of these events and the exploratory study of anticipated threats to water treatment barriers and distribution systems are important first responses in applied research. Alarming discoveries may then lead to exploratory research within the U.S. Environmental Protection Agency or, through federal research grants, to universities, foundations and private contractors. Once there is evidence of a health threat to water supplies, the best available technology is sought through field evaluation of treatment processes either in pilot-plant or full-scale operation.

Parallel to these investigations there is a need for risk assessment to ascertain who in the population may be compromised. Most often those at greatest risk include infants, senior citizens and special medical care groups (AIDS victims, patients with skin grafts and individuals on kidney dialysis). Most of these illnesses involve intestinal diseases, respiratory complications and, to a lesser extent, skin infections. Infections may be acquired from short term exposures to contaminated water supply — less than one day to as long as 30 days — before colonization of the pathogen results in disease. Adverse health effects from some chemical contaminants (THM's, lead, arsenic, pesticides, herbicides, radioactivity, etc.) may require continued exposures over 10 to 40 years.

At EPA, best available treatment is also evaluated on the basis of cost to achieve a benefit of avoiding one death among 10,000 or 100,000 consumers. At the same time, appropriate monitoring criteria are chosen and limits established within the sensitivity of detection methods. At that point, draft regulations are prepared with the assistance of agency scientists, university researchers and water supply professionals. The "straw man" rule (initial draft of a proposed rule) is then discussed in regional public hearings, revised and released for a public comment period. After the public comment period is closed, input is carefully reviewed and negotiated changes are made to finalize the rules. The new regulation is then published in the **Federal Register** with a date for national implementation. Thereafter, local utilities may obtain technical assistance related to rule changes and compliance issues from the federal and state agencies involved, and from various consulting engineering firms. Technical assistance areas of microbial interest include watershed management, filter instability, disinfectant concentration times, contact time (C$\bar{T}$) values for adequate inactivation of bacterial and viral pathogens, coliform biofilm occurrences, fecal contamination in distribution networks and waterborne outbreak investigations.

9 INFORMATION COLLECTION RULE (ICR)

During the initial development of recommendations for disinfectant/ disinfection by-product reductions it became apparent to an Advisory Committee (representatives from the water industry, State health agencies, environmental groups, consumer groups and EPA) that setting strict limits on the level of disinfectants applied could result in increased risk of pathogen breakthrough. What was needed was a substantial data base to show the effectiveness of microbial barriers in different treatment train configurations applied to a variety of raw source waters. No such data exists in full scale treatment operations; it would have to be generated from field operations. This need leads to the creation of the ICR in partnership with water utilities (Environmental Protection Agency, 1994). The most critical element of the ICR involves the collection of data on the concentrations of specific pathogens (*Crytosporidium, Giardia lamblia* and total culturable virus) in surface water used in the production of safe drinking water. Monitoring for selected pathogens and coliforms in the plant effluent will also be necessary to establish the percent removal through the treatment train configuration. This data base will be analyzed for the development of enhanced surface water treatment requirements and practical approaches to ground water disinfection by-product regulation.

The second element of the ICR will involve the collection of treatment plant operational data for characterizing disinfection by-product (DBP) formation and identifying surrogates for DBPs and DBP precursors. These data from the ICR will be used to: 1) characterize the source water constituents that influence DBP formation; 2) determine concentrations of DBPs in drinking water; 3) refine models for predicting DBP formation; and 4) establish cost-effective monitoring techniques.

The third element of the ICR will focus on bench- or pilot-scale studies on DBP precursor removal using either granular activated carbon or membrane filtration. The intent is to obtain more information on the cost effectiveness of these technologies for reducing DBP levels and to decrease the time utilities will need to install such technology.

The major issue with this strategy is the concern over the quality of the data that will be generated during the monitoring period. These data must be both accurate and precise in order to characterize the effectiveness of microbial barriers and the minimization of DBPs for different treatment-train configurations. Two major factors emerge: test sensitivity for selected pathogens and inter-laboratory capability to produce comparable data. This will prove to be no small task. For one example, protozoan test sensitivity may be only 30 percent and there is no certainty about cyst or oocyst viability. The only way to insure some control of laboratory performance will be to identify qualified laboratories that have been approved after an on-site visit and demonstration of satisfactory analysis of pathogen performance evaluation samples. This recent development has provided the opportunity for laboratories with expertise in the analysis of water samples for *Giardia, Crytosporidium* or viruses to receive special approval for testing water samples, provided they can demonstrate this expertise on unknown samples and pass an on-site review of their procedures (Environmental Protection Agency, 1994).

The burden for much of the microbial monitoring will be done by selected utilities serving over 100,000 customers. These utilities can be characterized as producing water supply from surface water or ground water systems under the direct influence of surface water. Source water will be monitored for 18 consecutive months on a monthly bases for: *Giardia, Cryptosporidium*, total culturable enteric viruses, total coliforms, and fecal coliforms or *E. coli*. Detection of ten or more pathogens per liter in the source water (100 liters) will require additional testing of the finished water (1,000 liters) for all of the identified pathogenic agents. Under consideration is the inclusion of *Clostridium perfringens* and coliphage (both somatic and F specific) to serve as surrogate indicators for protozoan pathogens and viruses, respectively. Particle size counts, in several size ranges, will also be performed to insure filter- removal efficiencies of particles and protozoans. The contribution to the data base from other systems would be tapered in intensity to providing information only on certain watershed characteristics, treatment processes used, sanitary surveys, coliform data and DBP precursor concentrations.

Missing from the ICR process are the public health aspects of the study. No provisions are currently included in the rule to collect data on health effects through surveillance and investigation of cases of *Cryptosporidium* in water supply. The ICR assumes that all oocysts detected in water are viable and potentially infectious and that all humans are equally susceptible to infection with a given inoculum. This does not appear to be part of the real world situation since *Cryptosporidium* can be detected in 30-50 percent of filtered and treated surface water sources. Any occurrences found in the finished water during the ICR study will place local and state health agencies in the difficult position of making a decision to issue a boil-water order that could unjustifiably undermine public confidence in their water supply. Perhaps the reason why there are few waterborne outbreaks caused by *Cryptosporidium* may be due, in part, to oocyst predators in the water flora. This biological controlling factor needs to be substantiated in field research.

10 PERSPECTIVE

Microbial regulations are changing as a result of more evidence of waterborne agents passing through existing microbial barriers. The search has started for more definitive information on treatment effectiveness for specific pathogens using conventional indicators and new surrogate candidates so that regulations for safe water supplies may be enhanced.

The process of improving public water safety through revised regulations has drawn water utility professionals, and State and Federal water authorities into a partnership of shared involvement. The tasks: better risk assessment and more responsive treatment to reduce adverse chemical and microbial exposures are further complicated by trade-offs in water supply processing that may adversely impact of microbial treatment barrier effectiveness. Reconfiguring water treatment practices and reinventing microbial regulations may lead to safer public water supplies.

References

AWWA Committee on Bacteriological Sampling Frequency, Committee Report: Current Practice in Bacteriological Sampling Frequency, Jour.Amer.Water Works Assoc.,1985;**77**:75-81.

Berger, P.S. and Y. Argaman. Assessment of Microbiology and Turbidity Standards for Drinking Water, U.S.Environmental Protection Agency,1983; EPA 570-9-83-001,Washington, D.C.

Committee on the Challenges of Modern Society (NATO/CCMS). **Drinking Water Microbiology.** NATO/CCMS Drinking Water Pilot Project Series CCMS128, U.S. Environmental Protection Agency,1984; EPA 570/9-84-006, Washington, D.C.

Craun, G.F. Statistics of waterborne outbreaks in the U.S. (1920-80). In: **Waterborne Diseases in the United States**, G.F. Craun (ed.), CRC Press, Inc., Boca Raton, Fl.1985.

Council of the European Communities. Proposal for a Council Directive Relating to the Quality of Water for Human Consumption, Official Jour. European Communities, 1975. **18**, No. C214/2.

Department of Health and Social Security. **The Bacteriological Examination of Water Supplies**. Reports on Public Health and Medical Subjects,no.71,4th ed. 1969, Ministry of Health, HMSO, London.

Donner, R.G. Seattle's Experience with distribution system sampling. Jour.Amer.Water Works Assoc.,1987;**79**:(11)38-41.

Environmental Protection Agency. National Interim Primary Drinking Water Regulations.1976. EPA-570/9-76-003. Office of Water Supply, Washington, D.C.

Environmental Protection Agency. Drinking Water; National Primary Drinking Water Regulations; Total Coliforms (Including Fecal Coliforms and E. coli); Final Rule. Federal Register **40 CFR Parts 141 and 142**, June 29, 1989, Washington, D.C.

Environmental Protection Agency. National Primary Drinking Water Regulations: Monitoring Requirements for Public Drinking Water Supplies: *Cryptosporidium, Giardia,* Viruses, Disinfection Byproducts, Water Treatment Plant Data and Other Information Requirements: Proposed Rule. Federal Register **59**:6332-6444, February 10,1994. Washington DC.

Geldreich, E.E. Status of Bacteriological Procedures Used by State and Municipal Laboratories for Potable Water Examination. Health Lab. Sci.1967;**4**:9-16.

Geldreich, E.E. and Rice,E.W. Occurrence, Significance, and Detection of *Klebsiella* in Water Systems. Jour. Amer. Water Works Assoc., 1987;**79**:74-80.

Hoskins, J.K. Revising the U.S. standards for drinking water quality: some considerations in the revision. Jour.Amer. Water Works Assoc., 1941;**33**:1804-1831.

Laboratory Certification Program Revision Committee. **Manual for the Certification of Laboratories Analyzing Drinking Water**.1991, EPA 570/9-90-008A U.S.Environmental Protection Agency, Washington D.C.

Lippy, E.C. and Waltrip, S.C. Waterborne disease outbreaks. 1946-1980: A thirty-five-year perspective. Jour. Amer. Water Works Assoc.1984;**76**;60-67.

Ludwig, F., Cocco, A., Edberg, S., Jarema, R. *et al.* Detection of Elevated Levels of Coliform Bacteria in a Public Water Supply - Connecticut. Morbidity and Mortality Weekly Report 1985;**34**:142-144.

McFeters, G.A.,J.S. Kippin and M.W. LeChevallier. Injured Coliforms in Drinking Water. Appl.Environ.Microbiol.1986; **51**:1-5.

Ministry of National Health and Welfare. **Microbiological Quality of Drinking Water**. Health and Welfare, Ottawa, Canada. 1977.

Payment. P. *et al.* A Randomized Trial to Evaluate the Risk of Gastrointestinal Disease due to Consumption of Drinking Water Meeting Current Microbiological Standards. Amer. Jour. Pub. Health 1991;**81**:703-708.

Safe Drinking Water Committee. **Drinking Water and Health**, Vol. 1, National Academy of Sciences, Washington, D.C. 1977.

Technical Subcommittee. Manual of recommended water sanitation practice accompanying United States Public Health Service Drinking Water Standards, 1942. Jour.Amer.Water Works Assoc.,1943. **35**:135-188.

World Health Organization. **European Standards for Drinking Water**, World Health Organization, Geneva, Switzerland 1970.

World Health Organization. **International Standards for Drinking Water**. World Health Organization, Geneva, Switzerland.   1971.

World Health Organization. **Guidelines for Drinking Water Quality**, Vol. 1, World Health Organization, Geneva, Switzerland.1984.

Pathogenic *E. coli*

Chapter 25
Effects Of Aquatic Environmental Stress On Enteric Bacterial Pathogens And Coliforms

Gordon A. McFeters

Department of Microbiology & Center for Biofilm Engineering Montana State University, Bozeman, Montana 59717, USA

1 INTRODUCTION

When enteric bacteria are exposed to sublethal levels of acute antibacterial agents and conditions, phenotypes of these bacteria become altered. Reports going back over four decades describe the failure of *Escherichia coli* to form colonies on commonly accepted media and incubation conditions following exposure to compounds such as phenolic antiseptics (Hershey, 1939) and chlorine (Milbauer and Grossowicz, 1959). These early observations of suboptimal bacterial recovery following exposure to a range of sublethal environmental stressors were interpreted as a form of cellular damage that reduced the culturability of allochthonous bacteria using selective methods that were not restrictive for freshly cultivated bacteria. An additional facet of the injury phenomenon that was also recognized relatively early (Harris, 1963) is the capability of the damaged cells to regain their more robust phenotype through a "revival" process where the damage is repaired under favorable circumstances. This recovery typically requires between one and three hours incubation under nonrestrictive conditions. During this time the level of colony forming units on selective media increase to the concentration seen using nonselective media. As pointed out by Litsky (1979), environmental conditions that stress enteric bacteria are a common feature of most aquatic systems and it should not be surprising that damaged cells are incapable of colony formation on many harsh selective media that continue to be used in microbiological analyses. Litsky further argued that this scenario represents an error among the early environmental microbiologists since many analytical approaches were borrowed, without modification, from medical applications where the target bacteria are both more numerous and directly isolated from an environment within a patient or animal to which the bacteria are well adapted.

Bacterial injury has been widely recognized for many years within the context of indicator organisms used to determine the microbiological quality of foods. As in many other environments, enteric bacteria associated with most food products are exposed to potentially damaging chemical and physical conditions during processing and preservation that result in cellular debilitation. This circumstance is important in the food industry as it can lead to an overly optimistic estimation of the safety of the product because a high proportion of the sublethally injured indicator bacteria will not be detected while pathogens, such as spore forming bacteria and viruses, might persist in high numbers as reviewed by Ray (1993). As in the case of many processed foods, the microbiological quality of treated water and wastewater is determined by the presence of enteric indicator bacteria that are, likewise, not well adapted to the ambient conditions in water where stressors including disinfectants are often present in low concentrations. In particular, the optimal concentration of disinfectant added as the treated water enters distribution systems rapidly decreases with flow time and distance due to reaction with biofilm components lining the pipes as well as with other materials. Therefore, it is not surprising that an average of 95% of the coliforms detected in three operating drinking water distribution systems in New England were injured and incapable of colony formation on the medium that was commonly used to determine water potability (McFeters, *et al.*, 1986). These and other similar observations from drinking water environments have been reviewed (McFeters, 1989; McFeters; 1990) and other reports have discussed the evolution of methods for the detection of injured coliform bacteria in water (LeChevallier and McFeters, 1984 & 1985). It has also been demonstrated that injured bacteria penetrate drinking water treatment plants and are not detected using accepted methods, then enter municipal distribution systems (Bucklin *et al.*, 1991). These observations collectively suggest that indicator organisms used to determine water contamination become unculturable, using accepted selective media and methods, following exposure to stressors such as disinfectants in drinking water. This occurrence brings into question the reliability of many of the methods that are commonly used in the assessment of water potability and safety, as in the case of foods.

The purpose of this chapter is to provide a ˙ brief overview of studies describing injured enteric bacteria following sublethal injury in aquatic environments. The significance of injured enteric bacteria in disinfected water systems will also be discussed. It is not intended that the coverage of the literature will be all-inclusive and the reader is directed to more complete reviews for additional details (McFeters, 1990; Singh & McFeters, 1990) .

2 INJURED BACTERIA IN AQUATIC SYSTEMS

Early evidence suggested the existence of injured bacteria within aquatic systems (Shipe and Cameron, 1952; Geldreich *et al.*,1987). The subsequent development and commercialization of a medium, designated m-T7, to specifically detect injured coliforms (LeChevallier *et al.*, 1983; LeChevallier *et al.*, 1984) in addition to the inclusion of a discussion of this and related methods in <u>Standard Methods for the Examination of Water and Wastewater</u> (APHA, 1992) made available techniques to test water for injured indicator bacteria. In addition, chlorine stressed coliforms and *E. coli* have been detected with greater efficiency using ColisureTM than with accepted reference methods (McFeters *et al.*, 1995).

A survey of over 200 potable water samples in Montana and Massachusetts revealed that greater than 50% of the coliforms present were injured (LeChevallier and McFeters, 1985). A subsequent study was then conducted to survey disinfected water from several cities in northeastern U.S. using m-T7 and m-Endo media (McFeters *et al.*, 1986). Coliforms found in 71 samples of water from distribution systems and 46 samples of water leaving treatment plants were 97% (mean value) injured. Similarly high levels of injured coliforms were also found in backwash water from filters used in the treatment of drinking water and in samples following the repair of a broken pipe within the distribution system. As reported elsewhere (LeChevallier *et al.*, 1983), injury levels were high in most of the chlorinated systems while source and cistern water contained fewer injured coliforms. Again, the fraction of the chlorinated water samples in which coliforms were detected with m-T7 medium but not by m-Endo, the medium that continues to be widely used for monitoring treated drinking water systems in the U.S., is striking. Bucklin *et al.* (1991) also demonstrated the penetration of significant levels of injured coliforms, that were undetected with m-Endo, through potable water treatment facilities. Therefore, established media used to determine water potability significantly underestimate the level of coliforms in disinfected water since injured bacteria become unculturable on commonly used selective media. These results underscore the significance of injured planktonic bacteria as a more sensitive indicator in the microbiological assessment of disinfected water and wastewater.

3 STRESSORS IN AQUATIC ENVIRONMENTS

A variety of chemical and physical stimuli have been implicated as causative factors in bacterial injury. Chlorine is the most commonly used biocide for the disinfection of potable water and wastewater streams and an early report by Mudge and Smith (1935) described a reversible form of bacterial inactivation in chlorinated systems. Later, McFeters and Camper (1983) demonstrated the very rapid occurrence of the injury process in coliforms using hypochlorous acid under controlled laboratory conditions. That was also observed in fecal streptococci by Lin (1976). The increasing use of chloramination as a disinfection strategy for the treatment of drinking water prompted a study demonstrating that exposure of enteric bacteria to monochloramine for ten minutes also resulted in a high levels (ie.>90%) of injury (Watters *et al.*, 1989). In another study carried out within operating municipal drinking water treatment facilities, the majority of bacteria isolated from the treated water were injured both with and without chlorination (Bucklin *et al.*, 1991). These findings suggest that factors in addition to disinfectants can lead to significant levels of injury in operating potable water systems. Early studies in our laboratory also suggested that influences other than disinfectants caused bacterial injury since suspensions of enteric bacteria exposed to natural stream water from different sources yielded reproducible levels of injury in the absence of added biocides (Bissonnette *et al.*, 1975). Alternative antimicrobial agents, such as ozone and chlorine dioxide, are being increasingly used to disinfect water. Of these, ozone has been shown to cause a reversible form of injury in *E. coli* that is manifest by the loss of culturability on m-FC medium that is accentuated at 44.5°C (Finch *et al.*, 1987). Additional studies are needed to determine if other such agents also elicit bacterial injury. Other potential sources of sublethal bacterial stress in the environment include metals, UV radiation, acidic pH, temperature extremes and biological factors. Laboratory studies determined that low concentrations of copper, typical of those found in potable water systems in the U.S., caused stress and an injury-concentration relationship was established (Domek, 1984). Wortman and Bissonnette (1985) further demonstrated that coliforms exposed to acid mine water became injured and were more sensitive to media containing deoxycholate. The possibility of biological interactions causing injury was also considered in aquatic systems since it is well established that high levels of heterotrophic count (HPC) bacteria suppress the detection of indicator bacteria in samples of drinking water (Geldreich *et al.*, 1972). Subsequent laboratory studies revealed that when pseudomonads exceeded coliforms by a factor of 105, greater than 50% of the *E. coli* became injured and were reversibly unculturable on selective media (LeChevallier and McFeters, 1985). These reports collectively indicate that a range of stressors that are ambient in many aquatic environments at sublethal levels can elicit injury and the reduced culturability of allochthonous bacteria when accepted selective media are used.

4 INJURY AND ENTERIC BACTERIAL PATHOGENS IN AQUATIC ENVIRONMENTS

Like coliforms, pathogenic bacteria that are transmitted via the waterborne route are susceptible to injury. In experiments done in our laboratory under controlled conditions, pathogens including *Yersinia enterocolitica, Salmonella typhimurium* and *Shigella* spp, were less susceptible to chlorine injury than the non-pathogenic coliforms and an enterotoxigenic strain (ETEC) of *E. coli* (LeChevallier *et al.*, 1985a). In addition, the virulence of these stressed populations was measured by the observation of 50% lethality levels when the bacteria were intraperitoneally injected into mice. The results of this study indicated that the virulence of the chlorine-stressed population was virtually eliminated while viability was retained. In the case of *Y. enterocolitica*, the loss of virulence associated with chlorine-induced injury resulted from the acquired inability to invade host cells while *S. typhimurium* and ETEC became unable to attach to appropriate cell cultures. A parallel set of experiments with copper-induced stress yielded similar results (Singh *et al.*, 1986; Singh & McFeters, 1987). Walsh and Bissonnette (1987) also demonstrated that sublethal concentrations of chlorine caused damage to the surface adhesins of ETEC and reduced their ability to attach to human leukocytes in an in-vitro assay. However, the death and injury of some enteropathogenic bacteria like *Campylobacter jejuni* (Blaser *et al.*, 1986) are seen at lower concentrations of chlorine. Terzieva and McFeters (1992) demonstrated that sublethal chlorine exposure results in the reversible loss of virulence-related properties of that organism, including attachment to, and invasion of cultured HeLa cells. These are representative of findings demonstrating that stressors such as chlorine, reversibly affect different determinants involved in the virulence phenotype. Another set of experiments was done to follow the *in vivo* revival, growth and pathogenicity of ETEC following chlorine-induced injury (Singh *et al.*, 1986). Following injection into ligated ileal loops in mice, injured ETEC suspensions recovered within 4 hours, with no increase in numbers. The same results were observed *in vitro* when chlorine-injured bacteria were incubated in saline containing homogenate of mouse intestinal mucosa but not in its absence. The enterotoxigenicity of injured ETEC cells was also determined, by injection into ligated ileal loops in rabbits, to be undiminished. These and similar results from others (Walsh and Bissonnette, 1987) indicate that chlorine-injured ETEC can recover both *in vitro* and within the gut and that it's toxigenic potential is retained. Singh and McFeters (1987) then demonstrated that when chlorine-injured ETEC were exposed to the additional stress of normal gastric acidity, in orogastrically inoculated mice, virulence remained undiminished in the gut. Therefore, injury resulting from exposure to chlorine at levels approximating those found in drinking water reversibly reduces the enteropathogenic potential of the pathogens examined and their virulence is recoverable within the mammalian gut following ingestion through the stomach. The reversible loss of virulence in the pathogens might be considered analogous to the reversible loss of culturability in indicator bacteria following sublethal exposure to chlorine. In addition, these results are similar, in some respects, to the viable but nonculturable state when a range of bacterial pathogens were exposed to simulated aquatic and marine environmental systems (Xu *et al.*, 1982; Roszak *et al.*, 1984). These findings provide additional evidence that allochthonous enteropathogenic bacteria retain both their viability and pathogenicity within various stressful aquatic environments.

5 SIGNIFICANCE OF INJURED BACTERIA IN WATER

Having established the occurrence of injured bacteria in aquatic environments, their importance remains to be discussed. From one perspective, detection of the entire population of viable indicator bacteria, including injured cells, provides an increased margin of analytical sensitivity and safety in the early detection of emerging microbiological problems within a system. That kind of information would allow the initiation of remedial action at an earlier stage and before the legally mandated limits are exceeded, triggering notification of both regulatory agencies and the public that the water might constitute a health hazard. For example, in Hopewell, VA (USA), the occurrence of coliform bacteria was sporadic in the potable water using m-Endo LES medium and the application of elevated chlorine levels appeared to resolve the problem (LeChevallier, unpublished results). However, the occurrence persisted until the system started using monochloramine. In this entire episode that spanned parts of three years, the detection of injured bacteria using m-T7 medium provided a much more accurate indication of coliform bacteria within the system. In addition, injury might explain outbreaks of waterborne morbidity where coliforms are either undetected or detected in low numbers. Further more, the penetration of injured coliforms through domestic water treatment facilities, as reported by Bucklin *et al.*, (1991), suggests that these undetected bacteria might inoculate the distribution system and set the stage for a subsequent regrowth event when conditions become suitable.

Published information available on the response of enteropathogenic bacteria to suboptimal disinfection in water also supports the importance of injured indicator bacteria in water. As discussed earlier, injured enteric pathogens regain virulence-related properties and continue to pose a health risk following ingestion into the mammalian gut. That relationship has also been demonstrated in the case of viable but nonculturable bacteria (Colwell *et al.*, 1985). It is also noteworthy that some enteric pathogens have been shown to be less sensitive to disinfectant-induced injury than coliform bacteria, as discussed earlier. Therefore, the detection of injured coliforms provides an additional degree of sensitivity in the detection of waterborne pathogenic bacteria and organisms that have classically been regarded as significant, from the perspective of public health, are likewise meaningful when injured. This greater analytical sensitivity could also be useful in epidemiological investigations of waterborne disease outbreaks where classical methods have failed to detect indicator bacteria.

6 SUMMARY

The information presented in this review supports the position that injured bacteria are important when considering the microbial ecology and public health microbiology of a wide range of aquatic systems. For example, ambient environmental conditions within operating potable water systems, including suboptimal conditions of disinfection, result in the occurrence of injured bacteria. Under some circumstances these stressed subpopulations, which are not culturable when using most accepted methods for water quality monitoring, can represent the majority of indicator bacteria within a system. Such an occurrence suggests that more sensitive and representative water quality information would be obtained if media or strategies were used to detect the sublethally injured subpopulation of indicator bacteria. Such information has been demonstrated to be of value in the detection of operational and mechanical problems within water treatment and distribution systems. Hence, the detection of injured bacteria affords an added measure of accuracy to assist in the early recognition of treatment deficiencies or contamination within operating water systems. Continuing efforts are needed to improve the detection of injured and nonculturable bacteria. More importantly, the adoption of methods to enhance the detection of injured bacteria within the drinking water industry as well as the support of this concept by appropriate regulatory agencies would help ensure the safety of potable water supplies. The persistent worldwide incidence of waterborne morbidity underscores the need for more sensitive and accurate indicators of potential microbiological health hazards in water.

Acknowledgements

I am grateful to Dr. Mark LeChevallier and Ms. Cheryl Norton for collecting the data from Hopewell, VA, and to Dr. Barry H. Pyle for prepublication review of this manuscript as well as the many colleagues and former students who collaborated in the work reported here. We also acknowledge current support from the National Aeronautics and Space Administration, the National Science Foundation /Office of Polar Programs and the Center for Biofilm Engineering at Montana State University (cooperative agreement EEC-8907039).

References

APHA, AWWA and WEF. 9212 Stressed organisms. In Standard Methods for the Examination of Water and Wastewater, ed. A.E. Greenberg, 1992;PP. 924-926. APHA, Washington.

Bissonnette GK, Jezeski JJ, McFeters GA and Stuart DG. Influence of environmental stress on enumeration of indicator bacteria from natural waters. Appl. Microbiol. 1975;29:186-194.

Blaser MJ, Smith PF, Wang W-LL and Hoff JC. Inactivation of C. jejuni by chlorine and monochloramine. Appl. Environ. Microbiol. 1986;51:307-311.

Bucklin KE, McFeters GA and Amirtharaja A. Penetration of coliforms through municipal drinking water filters. Water Res. 1991;25:1013-1017.

Colwell RR, Brayton PR, Grimes DJ and Roszak DB. Viable but non-recoverable V. cholerae and related pathogens in the environment: implications for release of genetically engineered microorganisms. Bio/Technol. 1985;3:817-820.

Domek MJ, LeChevallier MW, Cameron SC and McFeters GA. Evidence for the role of copper in the injury process of coliforms in drinking water. Appl. Environ. Microbiol. 1984;48:289-293.

Finch GR, Stiles ME and Smith DW. Recovery of a marker strain of E. coli from ozonated water by membrane filtration. Appl. Environ. Microbiol. 1987;53:2894-2896.

Geldreich EE, Allen MJ and Taylor RH. Interferences to coliform detection in potable water supplies. In Evaluation of the Microbiology Standards for Drinking Water, ed. CW. Hendricks, 1987;pp. 13-20. USEPA (EPA-570/9-78-00C), Washington.

Geldreich EE, Nash MD, Reasoner DJ and Taylor RH. The necessity of controlling bacterial populations in potable waters: community water supply. J. Amer. Water Works Assoc. 1972;64:596-602.

Harris ND. The influence of recovery medium and the incubation temperature on the survival of damaged bacteria. J. Appl. Bacteriol. 1963;26:387-397.

Hershey AD. Factors limiting bacterial growth: properties of E. coli surviving sublethal temperatures. J. Bacteriol. 1939;38:563-579.

LeChevallier MW, Cameron SC and McFeters GA. New medium for the recovery of coliform bacteria from drinking water. Appl. Environ. Microbiol. 1983;45:484-492.

LeChevallier MW, Jakanoski PE and McFeters GA. Evaluation of m-T7 agar as a fecal coliform medium. Appl. Environ. Microbiol. 1983;48:371-375.

LeChevallier MW and McFeters GA. Recent advances in coliform methodology. J. Environ. Health 1984;47:5-9.

LeChevallier MW and McFeters GA. Enumerating injured coliforms in drinking water. J. Amer. Water Works Assoc. 1985;77:81-87.

LeChevallier MW, Singh A, Schiemann DA and McFeters GA. Changes in virulence of waterborne enteropathogens with chlorine injury. Appl. Environ. Microbiol. 1985a;50:412-419.

Lin S. Membrane filter method for recovery of coliforms in chlorinated sewage effluents. Appl. Environ. Microbiol. 1976;32:547-552.

Litsky W. Gut critters are stressed in the environment, more stressed by isolation procedures. In Aquatic Microbial Ecology, eds. R.R. Colwell and J. Foster, 1979;pp. 345-347. A Maryland Sea Grant Publication, College Park.

McFeters GA and Camper AK. Enumeration of indicator bacteria exposed to chlorine. In Advances in Applied Microbiol, vol. 29, ed. A.I. Laskin, 1983;pp. 177-193. Academic Press, New York.

McFeters GA Detection and significance of injured indicator and pathogenic bacteria in water. In Injured Index and Pathogenic Bacteria: Occurrence and Detection in Foods, Water and Feeds, ed. B. Ray, 1989;pp. 179-210. CRC Press, Boca Raton.

McFeters GA Enumeration, occurrence and significance of injured indicator bacteria in drinking water. In Drinking Water Microbiology: Progress and Recent Developments, ed. G.A. McFeters, 1990;pp. 478-492. Springer-Verlag, New York.

McFeters GA, Kippin JS and LeChevallier MW. Injured coliforms in drinking water. Appl. Environ. Microbiol. 1986;51:1-5.

McFeters GA, Broadaway SC, Pyle BH and Pickett M. Equivalency of Colisure and accepted methods for detection of chlorine injured coliforms and *E. coli*. 1995;(in this volume)

Mudge CS and Smith FR. Relation of action of chlorine to bacterial death. Amer. J. Public Health 1935;25:442-447.

Milbauer R and Grossowicz N. Reactivation of chlorine-inactivated *E. coli*. Appl. Microbiol. 1959;7:67-70.

Ray B. Sublethal injury, bacteriocins and food microbiology. Amer. Soc. Microbiol. News 1993;59:285-291.

Roszak DB, Grimes DJ and Colwell RR. Viable but nonrecoverable stage of S. enteritidis in aquatic systems. Can. J. Microbiol. 1984;30:334-338.

Shipe EL and GM Cameron. A comparison of the membrane filter with most probable number method for coliform determination from several waters. Appl. Microbiol. 1952;2:85-87.

Singh A , Yeager R and McFeters GA. Assessment of in vivo revival, growth and pathogenicity of *E. coli* strains after copper and chlorine injury. Appl. Environ. Microbiol. 1986;52:832-837.

Singh A and McFeters GA. Survival and virulence of copper- and chlorine-stressed Y. enterocolitica in experimentally infected mice. Appl. Environ. Microbiol. 1987;53:1768-1774.

Singh A and McFeters GA. Injury of enteropathogenic bacteria in drinking water. In Drinking Water Microbiology: Progress and Recent Developments, ed. G.A. McFeters, 1990;pp. 368-379. Springer-Verlag, New York.

Terzieva SI and McFeters GA. Effect of chlorine on some virulence-related properties of C. jejuni. Int. J. Environ. Health Res. 1992;2:24-32.

Walsh SM and Bissonnette GK. Effect of chlorine injury on heat-labile enterotoxin production in enterotoxigenic *E. coli*. Can. J. Microbiol. 1987;33:1091-1096.

Watters SK, Pyle BH, LeChevallier MW and McFeters GA. Enumeration of E. cloacae after chloramine exposure. Appl. Environ. Microbiol. 1989;55:3226-3228.

Wortman AT and Bissonnette GK. Injury and repair of *E. coli* damaged by exposure to acid mine water. Water Res. 1985;19:1291-1297.

Xu H, Roberts N, Singleton SL, Attwell RW, Grimes DJ and Colwell RR. Survival and viability of nonculturable *E. coli* and V. cholerae in estuarine and marine environments. Microb. Ecology 1982;8:313-323.

Chapter 26
Biochemical Studies On Enterohaemorrhagic *Escherichia Coli* (EHEC)

Karl A. Bettleheim

Biomedical Reference Laboratory. Victorian Infectious Diseases Reference Laboratory, Fairfield Hospital, Fairfield, Victoria 3078, Australia

1 INTRODUCTION

For many years strains of *Escherichia coli* have been differentiated by means of biotyping, making use of the fact that strains of *E. coli* vary in their ability to ferment certain carbohydrates and decarboxylate amino acids. A system was developed in 1969 as part of a study on isolates from cases of chronic urinary tract infections [1], which used the fermentation reactions of 16 carbohydrates or carbohydrate derivatives. It was found that within the internationally accepted system of serotyping, biotyping provided a useful adjunct to differentiate related strains. Subsequently in a series of studies on the colonization of neonates by *E. coli,* biotyping was again found to be very valuable as a means of differentiating strains and determining their ecology [2,3]. Although it has been shown that fermentation reactions can vary as a result of the presence of extrachromosomal elements, this did not invalidate the use of biotyping [4]. A series of studies by Crichton and Old [5,6,7] have elegantly demonstrated the usefulness of biotyping as a means of differentiating strains of *E. coli.*

A number of commercial systems have been developed for the identification of microorganisms. Using the MicroScan dried conventional gram-negative identification panel (Baxter Diagnostics, Inc., West Sacramento, CA, U.S.A.), it was shown that strains of *E. coli* belonging to serogroup O157 [8] produced a very limited number of profiles. The majority of Enterohaemorrhagic *E. coli* (EHEC) of serotype O157.H7 tested were restricted to one profile, which was not found among other *E. coli.*

Another typing system developed over the last few years uses strips of reagent coated plastic wells, like microtitre trays [9] (Microbact System, Disposable Products Pty, Ltd. Technology Park, SA, Australia), which are inoculated with a suspension of the organisms under test. This system purely requires the test strips and few extra reagents, it is very convenient to use for the rapid screening of single strains. In this study strains of EHEC O157.H7 and O157.H- of both clinical and environmental origin were examined using this system. In addition, strains of EHEC belonging to other serotypes as well as non-EHEC belonging to a variety of serotypes including ones commonly associated with EHEC were examined.

As part of an investigation into the haemolysins of *E. coli* Beutin *et al.* (1989) [10] noted a close association between the production of Shiga-like toxin(s) and enterohaemolysin. Of EHEC strains belonging to serogroups O157, as well as O26, O111, and O116, they found that of 64 EHEC, 60 (93.8%) were positive for the production of enterohaemolysin. These observations have since been confirmed with Australian strains [1]. These investigations were extended with the particular aim to determine if this applied to EHEC in general and if there is a unique biochemical characteristic of EHEC with respect to enterohaemolysin production and biochemical reactions. This could be used for the identification of EHEC from patient and environmental specimens.

2 MATERIALS AND METHODS

2.1 Bacterial Strains

The strains *E. coli* tested were from the culture collection of the Biomedical Reference Laboratory, Victorian Infectious Diseases Reference Laboratory, Fairfield Hospital. EHEC belonging to serotypes including O157.H7, O157.H- as well as strains belonging to serogroups O26, O48, O81, O91, O111, O113 and O128 were included. While many were from various regions of Australia, others were obtained from Germany (Dr. L. Beutin, Robert Koch Institut, Berlin, Germany), U.S.A. (Dr. N. Strockbine, CDC, Atlanta, GA, U.S.A.) and Africa (Dr. R. Robins-Browne, Royal Children's Hospital, Melbourne, Australia). Also some non-O157 EHEC from New Zealand were included in this study (Dr. H. Brooks, University of Otago, Dunedin, New Zealand). Non-EHEC included strains from cases of travellers' diarrhoea, cases of urinary tract infections, and isolates from healthy Australian children and healthy German children (Dr. Beutin, Robert Koch Institut, Berlin, Germany). Also included were Enteropathogenic *E. coli* (EPEC) strains from described outbreaks in the past in the United Kingdom [12,13].

2.2 Microbiological Studies

All the strains were tested by the Microbact System, following the manufacturers, instructions. The tests provided on the test strips are the following: Decarboxylation of Lysine, Ornithine and Arginine; H_2S, Production of Urease and Indole; Fermentation with acid production of Glucose, Mannitol, Xylose, Inositol, Sorbitol, Rhamnose, Sucrose, Lactose, Arabinose, Adonitol, Raffinose and Salicin; Hydrolysis of o-Nitrophenyl-β-D-galactopyranoside (ONPG); Voges-Proskauer Test (VP); Utilization of Citrate; Production Indolepyruvate by deamination of Tryptophan (TDA); Gelatin Liquefaction; and Malonate Inhibition. Anaerobic conditions are applied to some tests by use of mineral oil and the strips sealed with plastic tape. Following overnight incubation the colour changes were recorded. The three tests requiring additional reagents (Indole production, TDA and VP) were performed as per the manufacturer' instructions. On the basis of the tests, a profile for the organism is obtained (Figure 26.1) which can be either manually compared with the list provided or generated by a computer programme, which is provided by the manufacturer.

The organisms were O and H serotyped [14,15]. They were tested for their ability to produce Verotoxins (VT) (Shiga-like toxins, SLT) [16,17] both by their cytopathic effects on Vero cells and by enzyme-linked immunosorbant assay. Their ability to produce the different haemolysins was determined on sheep-blood agar and washed sheep-blood agar [10,11]. The test for VT also demonstrated the production of Cytotoxic Necrotizing Factor (CNF). They were also tested for their ability to produce the heat-labile enterotoxin (LT) using the Phadebact system [18].

Reaction	Lysine	Ornithine	H2S	Glucose	Mannitol	Xylose	ONPG	Indole	Urease	V.P.	Citrate	TDA	Gelatin	Malonate	Inositol	Sorbitol	Rhamnose	Sucrose	Lactose	Arabinose	Adonitol	Raffinose	Salicin	Arginine
Reaction Index	4	2	1	4	2	1	4	2	1	4	2	1	4	2	1	4	2	1	4	2	1	4	2	1
Result	+	+	−	+	+	+	+	+	−	−	−	−	−	−	−	−	−	−	+	+	−	−	−	−
Profile		6			7			6			0			0			0			6			0	
Result	+	+	−	+	+	+	+	+	−	−	−	−	−	−	−	−	−	−	+	+	−	−	−	+
Profile		6			7			6			0			0			0			6			1	
Result	+	+	−	+	+	+	+	+	−	−	−	−	−	−	−	−	−	−	+	+	−	+	−	−
Profile		6			7			6			0			0			0			6			4	
Result	+	+	−	+	+	+	+	+	−	−	−	−	−	−	−	−	−	−	+	+	−	+	−	+
Profile		6			7			6			0			0			0			6			5	
Result	+	+	−	+	+	+	+	+	−	−	−	−	−	−	−	−	−	+	+	+	−	−	−	−
Profile		6			7			6			0			0			1			6			0	
Result	+	+	−	+	+	+	+	+	−	−	−	−	−	−	−	−	−	+	+	+	−	−	−	+
Profile		6			7			6			0			0			1			6			1	

Figure 26.1 *Determination of Microbact profiles*

3 RESULTS

Initially 16 strains were tested by the MicroScan system. These strains comprised 12 Australian toxigenic O157 isolates, two non-toxigenic O157 isolates and two toxigenic non-O157 isolates. The results obtained were similar to those described for the American studies [8]. These 16 strains were then tested by the Microbact 24E system and again it was shown that the 12 toxigenic O157 isolates gave a restricted set of profiles. It was therefore decided to proceed with a further selection of strains. The 141 strains of *E. coli* tested produced 45 different profiles on the Microbact system. The seven profiles, which were most commonly associated with O157 strains were characterised by poor fermentation ability. The areas of isolation of the strains included Australia, Germany, U.S.A. and Africa. Both EHEC O157 strains, which did not fall into those profiles were Australian environmental O157.H- isolates. One of the strains had produced SLT II on first isolation but lost this ability on subculture. All the SLT-producing strains of O157 produced enterohaemolysin except one Australian patient isolate of O157.H- which produced SLT II. Non-SLT producing strains belonging to serogroup O157 did not produce enterohaemolysin, if motile they did not produce the H7 antigen and were not isolated from situations suggestive of EHEC infections. They fell into a variety of profiles.

Of the non-O157 isolates six strains gave the profile, which was associated with most of the EHEC O157 strains. Three of these were also EHEC of serotypes O111.H- and O113.H21. This profile is characterised by the inability of the strains to ferment the following five carbohydrates: Sorbitol, Rhamnose, Sucrose, Raffinose and Salicin. Of the 47 SLT producing O157 strains, included in this study, 28 (59.6%) did not ferment any of these carbohydrates and 17 (36.1%) fermented only one of these substrates. Only two (4.2%) strains fermented three of the substrates. Thus 96% of SLT producing strains of *E. coli* O157 fermented none or or only one of these carbohydrates. In addition eight (72.7%) of 11 non-SLT producing strains of O157 fell into this poorly fermenting group. A number of these may have been SLT producers as exemplified by the fact that most were enterohaemolysin positive. Of the non-O157 strains 14 of 29 (48.3%) of SLT producing strains fermented one or none of these carbohydrates while only seven out of 54 (13.0%) of non SLT producers fermented one or less of these carbohydrates.

While the majority of the SLT producing O157 strains gave this limited fermentation profile, a far greater variety of profiles was noted among the other SLT-producing serotypes. Of the 16 O111 strains studied, there were five non-SLT and 11 SLT-producing strains. Most of them fermented rhamnose, only two of the 11 SLT-producing strains did not. However, of the remaining carbohydrates none of the strains fermented sucrose or salicin and four and one of 11 SLT-producing strains fermented sorbitol or raffinose respectively. Again the fermentation ability of the EHEC was less than that of the non-EHEC but this was not as marked as with the O157 strains. Simmilarly of the five SLT-producing O113 strains all but one fermented only one or none of these five carbohydrates.

With respect to some of the other EHEC serotypes there was a great variety of fermentative types observed, although the general tendency of reduced fermentative ability was noted.

Previous observations on the strong relationship between enterohaemolysin and SLT-production was noted. A notable exception was the SLT-producing O113 strains which all produced alpha-haemolysin. However, 46 (97.9%) out of 47 SLT-producing O157 strains were enterohaemolysin producers and 23 (79.3%) out of 29 SLT-producing non-O157 strains were enterohaemolysin producers. Five of these six were O113 strains suggesting that these may have a particular characteristic. The sixth was an environmental strain of serotype O81.H, which also fermented sorbitol, rhamnose and raffinose.

4 DISCUSSION

These observations indicate that the majority of the EHEC O157 strains can be grouped into a small number of related profiles. These profiles are characterised by the strains' inability to ferment a number of carbohydrate substrates including sorbitol, rhamnose, sucrose, raffinose and salicin and their inability to decarboxylate arginine. The majority of SLT producing *E. coli*, which were either O157.H- or O157.H7 from four continents (Africa, Australia, Europe and North America) fell into these groups, which were characterised by very limited fermentative ability. Only two environmental Australian O157.H- strains producing SLT gave other profiles. While the other SLT-producing serotypes gave a diverse range of profiles, there was still a tendency for less fermentative ability, although not as marked as with the O157 strains.

These studies suggest that most of the SLT-producing strains of *E. coli* O157.H7 and O157.H- may be descended from a single clone, which has spread throughout the world. With the use of the Microbact 24E system as described, it should be possible to identify these organisms from primary isolation rapidly as to whether they belong to that particular clone and thus are likely to be toxigenic. This should simplify screening for these organisms, both of envionmental as well as patient samples, particularly as this technique will in addition identify the organisms as *E. coli*. Used in conjunction with one of the rapid screening methods for O157 antigen which are now available it would be very advantageous for investigations of these newly emerging pathogens. If the organisms are also tested for their ability to produce enterohaemolysin, it could rapidly establish the strong possibility of a test strain being a potential EHEC.

On the basis of the evidence presented here, it is suggested that an *E. coli*, which produces enterohaemolysin and has a poor fermentative ability is likely to be an EHEC. From this observation selected colonies from a primary isolation could be tested for haemolysis on washed and unwashed sheep blood agar and on MacConkey agars in which the lactose is replaced by sorbitol, rhamnose, raffinose or sucrose. Strains which are haemolytic only on the washed sheep blood agar and only ferment maximally two of the carbohydrates would be then tested further. Such tests could include specific DNA probes for SLT specific sequences, serological tests for O157 or O111 or other specific tests depending on what is being specifically sought. Such a rapid screening could be a cheap and efficient first stage, particularly, when environmetal sampling such as natural waters, foods or food manufacturing processes or potential contacts of patients are being investigated. There is no microbiological screening procedure which will be 100% effective and neither is this one. Therefore, with all procedures caution in interpretation is important. There should also be the awareness that there will be some EHEC, which will be missed by these tests and there will be non-EHEC which will initially be included. These latter can be easily excluded by confirmatory tests. The problem of the former is somewhat greater. However, if there is a strong suspicion that EHEC are present in a sample under test than a competent microbiologist should be aware of the limitations of this series of screening tests.

References

1. Bettelheim K.A. and Taylor J. A study of *Escherichia coli* isolated from chronic urinary infection. *J. med. Microbiol.* **2**, 225-236 (1969).

2. Bettelheim K.A., Teoh-Chan C.H., Chandler M.E., O'Farrell S.M., Rahamin L., Shaw E.J. and Shooter R.A. Further studies of *Escherichia coli* in babies after normal delivery.*J. Hyg. Camb.* **73**, 277-285 (1974).

3. Bettelheim K.A., Teoh-Chan C.H., Chandler M.E., O'Farrell S.M., Rahamin L., Shaw E.J. and Shooter R.A. Spread of *Escherichia coli* colonizing newborn babies and their mothers.*J. Hyg. Camb.* **73**, 277-285 (1974).

4. Shinebaum R., Shaw E.J., Bettelheim K.A. and Dickerson A.G. Transfer of invertase production from a wild strain of *Escherichia coli*. *Zbl. Bakt. Hyg., I.Abt. Orig. A.* **237**, 189-195 (1977).

5. Crichton P.B. and Old D.C. Differentiation of strains of *Escherichia coli*: Multiple typing approach. *J. clin. Microbiol.* **11**, 635-640 (1980).

6. Crichton P.B. and Old D.C. A biotyping scheme for the subspecific discrimination of *Escherichia coli. J. med. Microbiol.* **15**, 233-242 (1982).

7. Crichton P.B. and Old D.C. Numerical index of the discriminatory ability of biotyping and resistotyping for strains of *Escherichia coli*: *Epidemiol. Infect.* **108**, 279-286 (1992).

8. Abbott S.L., Hanson D.F., Felland T.D., Connell S., Shum A.H. and Janda J.M. *Escherichia coli* O157:H7 generates a unique biochemical profile on MicroScan conventional Gram-negative identification panels. *J. clin. Microbiol.* **32**, 823-824 (1994).

9. Mugg P. and Hill A. Comparison of the Microbact 12E and 24E systems and the API 20E system for the identification of Enterobacteriaceae. *J. Hyg., Camb.* **87**, 287-297 (1981).

10. Beutin L., Montenegro M.A., Ørskov I., Ørskov F., Prada J., Zimmermann S. and Stephan R. Close association of verocytotoxin (Shiga-like Toxin) production with

enterohemolysin production in strains of *Escherichia coli*. *J. clin. Microbiol.* **27**, 2559-2564 (1989).

11   Bettelheim K.A. Identification of enterohaemorrhagic *Escherichia coli* by means of their production of enterohaemolysin. *J. Appl. Bact.* **79**, 178-180 (1995).

12.   Robins-Browne R.M., Yam W.C., O'Gorman L.E. and Bettelheim K.A. Examination of archetypal strains of enteropathogenic *Escherichia coli* for properties associated with bacterial virulence. *J. med. Microbiol.* **38**, 222-226.

13.   Bettelheim K.A., Drabu Y., O'Farrell S., Shaw E.J., Tabaqchali S. and Shooter R.A. Relationship of an epidemic strain of *Escherichia coli* O125.H21 to other serotypes of *E. coli* during an outbreak situation in a neonatal ward. *Zbl. Bakt. Hyg., I.Abt. Orig. A.* **253**, 509-514 (1983).

14.   Chandler M.E. and Bettelheim K.A. A rapid method of identifying *Escherichia coli* "H" antigens. *Zbl. Bakt. Hyg., I.Abt. Orig.A*, **129**, 74-79 (1974).

15.   Bettelheim K.A. and Thompson C.J. New Method of serotyping *Escherichia coli*: Implementation and Verification. *J. Clin. Microbiol.* **25**, 781-786 (1987).

16.   Konowalchuk J., Speirs J.L. and Stavric S. (1977) Vero response to a cytotoxin of *Escherichia coli*. *Infection and Immunity* **18**,775-779.

17.   Acheson D.W.K., Keusch G.T. Lightowlers M. and Donohue-Rolfe A. Enzyme-linked immunosorbent assay for Shiga toxin and Shiga-like toxin II using P_1 glycoprotein from hydatid cysts. *J. Infect. Dis.* **127**,1145-50 (1990) .

18.   Bettelheim K.A., Gracey M. and Wadström T. The use of the coagglutination test to determine whether Australian and New Zealand isolates of *Escherichia coli* produce the heat-labile enterotoxin. *Zbl. Bakt. Hyg. A*, **260**,293-296 (1985).

Chapter 27
The Operational Significance Of Coliform Species In Treated Water

J.G. O'Neill and O.T. Parry

Yorkshire Water Services Ltd,
Leeds LS1 4BG

1 BACKGROUND

In Yorkshire Water Services Ltd and it's predecessors, identification of coliforms to species level using a variety of systems has been used since 1976. For the past ten years the API 20E system (BioMerieux, France) has been used to identify all coliforms isolated from treated water. The reasons for identifying coliforms found in treated water to species level are twofold; firstly to provide information, as far as is possible on the origin and possible health risk presented by the organism, and secondly to provide information which may contribute towards identification of point source contamination or other problems for which remedial action can be put into place. Data and incidents selected from ten years of experience, which illustrate the value of the use of identifying coliforms to species level, are described below.

2 GROWTH V CONTAMINATION IN DISTRIBUTED WATER

Isolation of coliforms from treated and distributed water in the British water industry is becoming an increasingly rare event. For the industry overall in England and Wales in 1994, treated water samples with no coliforms as a percentage of those taken were: 99.9% ex-works, 99.7% from service reservoirs and 99.3% at customer's taps (1). However, as the situation improves it becomes increasingly necessary to investigate and explain each individual failure. One particular aspect of such investigations, in relation to coliforms detected at customer's taps, is to distinguish between contamination and biofilm growth. The most vulnerable point for contamination of distributed water is during storage in service reservoirs and this can be the origin of detection of coliforms at customer's taps. However, it is now well established that isolation of coliforms from customer's taps can also result from biofilm growth in the distribution pipework (2,3). Coliforms from biofilm are regarded with less significance than if from contamination after treatment. Identification of coliforms to species level can provide some insight into these processes. Table 1 lists the coliform species isolated in Yorkshire over a two year period (1993 and 1994) from service reservoirs and customer's taps. The total number of isolates in each case is approximately 200. Detection of coliforms leaving the treatment works must also be considered in any investigation but the incidence is too low to significantly affect this data.

It can clearly be seen from Table 27.1 that the species make-up is different between the

Table 27.1 *Percentage of total isolates for each coliform species isolated from routinely taken samples of service reservoirs and customer's taps in Yorkshire*

	SERVICE RESERVOIRS	CUSTOMER'S TAPS
Escherichia coli	30	9
Enterobacter intermedium	17	5
Citrobacter freundii	11	4
Klebsiella pneumoniae	10	24
Enterobacter agglomerans	9	15
Hafnia alvei	6	1
Enterobacter amnigenus	6	3
Enterobacter cloacae	5	20
Klebsiella oxytoca	5	11
Kluyvera species	2	2
Serratia species	-	4
Enterobacter species	-	5

two data sets. This difference in the species make-up between service reservoirs and customer's taps represents essentially, although not exclusively, different routes of access to treated water. Whereas the isolates from service reservoirs are more likely to be, for example, ingress through the roof, the isolates from customer's taps are more likely to be from biofilm growth either in the network or in customer's plumbing. During the period in question there was a relatively high proportion of *E.coli* at service reservoirs. This was usually found to originate from birds or small mammals such as rabbits and moles. Other coliforms in service reservoirs are proportionally likely to be of soil origin. The *E.coli* and some of the other coliforms tend to die off or become diluted out in the distribution system. Those organisms surviving or increasing in numbers in biofilms become more common into the distribution system. Data from customer's taps in other English water companies (2) and from the USA (3) is similar, with the predominating genera being *Klebsiella* and *Enterobacter*, which are regarded as primarily biofilm related. Conversely, in a Scottish study, *Serratia fonticola* was found to be the predominating species in both service reservoirs and at customer's taps (2). The reason for this is under investigation.

3 CLASSIFICATION OF OPERATIONAL PROBLEMS

One important aspect of separating coliform contamination at service reservoirs from biofilm growth as described above is to focus appropriate remedial action. Service reservoirs are commonly concrete structures and will slowly deteriorate. However, any indication of leakage into a service reservoir is likely to quickly result in remedial action e.g. roof repair. Detection of coliforms can often be the first indication of such problems.

Routine identification of coliforms to species level can be valuable in classifying operational problems in order to identify appropriate remedial action. As a specific example, a *Klebsiella oxytoca* in a treated water sample taken from a treatment works (in the presence of 0.5 mg/l chlorine) was found in a service reservoir and in distribution from this treatment works 2 days later. This association, particularly when repeated some weeks later, identified the situation as a single problem rather than two or even three different problems. Again, species identification in linked service reservoirs has indicated a single rather than multiple points of contamination with subsequent appropriate remedial action. It can also be valuable to link service reservoir and customer tap isolation of not only species but also biochemically identical strains (identical reference numbers with the API 20E system). Although the speciation of isolated coliforms does not contribute towards identification of the particular problem in every case, only a very small number of savings of unnecessary remedial work will more than cover additional laboratory costs.

A particular situation in which *Citrobacter freundii*, predominantly a soil organism, has been persistently isolated from distributed water has been observed on a number of occasions following transient contamination of a main with soil or water. Figure 1 shows a possible model for this persistence compared with contaminating material containing *E.coli*. Following a point source contamination, at any specific point in the network the *E.coli* declines rapidly primarily by washout as does the *C.freundii*. However, the *C.freundii* can often be isolated for several days before it falls below the level of detection. The interpretation of observations of this nature is that the *C.freundii* is temporarily incorporated into the biofilm with some growth and subsequent release. *C.freundii* is not usually a predominating biofilm organism as can be seen in Table 27.1. Retrospective examination of events following persistent *C.freundii* isolation have been linked with otherwise unremarked breaks in pipelines. This situation is somewhat different to a more general biofilm situation where any of the coliforms

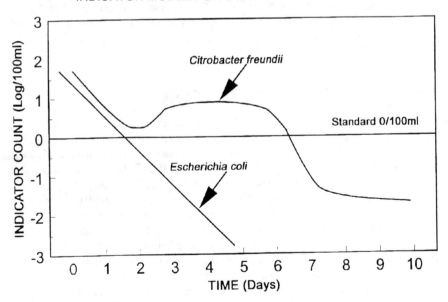

Figure 27.1 *Theoretical Model for development of numbers of E.coli and C.freundii following point source contamination of a distribution network.*

could be represented, but primarily *Klebsiella* and *Enterobacter* as indicated in Table 27.1.

As a further example, in a specific investigation of persistent coliforms in a network it was found that all the isolates were *E. amnigenus* and *Hafnia alvei*. All available evidence suggests that this was due to primary biofilm growth with long retention, low chlorine, and unusually high temperature but as these organisms are also relatively unusual in biofilm an unidentified contamination event as illustrated in Figure 27.1 cannot be entirely ruled out.

4 HEALTH EFFECTS

In a recent study (4), it was concluded that in the UK no outbreaks of infection have ever been associated with growth of heterotrophic bacteria, including coliforms, in drinking water. Much of the data examined in this study was derived from the use of API20E and similar systems. The API20E system was actually designed for medical purposes and although generally satisfactory in providing a species name, some isolates from the water habitat are indeterminate or can be given a spurious identification. A particularly dramatic example is an isolate which can be found occasionally in treated water for which the system provides a 'satisfactory' identification as *Yersinia pestis*, the causative agent of bubonic plague. On further investigation it has proved to be *Serratia fonticola*, a harmless organism found mainly in surface water (4). The latest version of the system (ATBPlus version 2.2.7) may have overcome this problem.

The essential historical use of the analysis of water for the coliform group of bacteria in health terms has been indirect, as indicators of faecal contamination. *Escherichia coli* has always been regarded as the unequivocal indicator of faecal material with no independent existence outside the body (5), and constitutes 97% of the coliform population of human faeces (6). The significance of other members of the coliform group is more ambiguous as outlined above. Most of the coliform group can be found in faeces in small numbers but are also widespread in soil and water. However, *Enterobacter intermedium*, *E. amnigenus* and *S. fonticola* are not found in faeces (5). *Citrobacter* is also not regarded as being of faecal origin (6).

5 LABORATORY AND SAMPLING CONTAMINATION

Perhaps the most valuable contribution of coliform identification in the Company over the ten year period has been identification of sampling and laboratory contamination due to localised growth of coliforms. Perhaps fortunately these have involved unusual organisms, at a time when the company had several microbiology laboratories. In the two examples of contamination error in the laboratory described below, around 25% of the coliforms found over a six month period from a particular laboratory were a single species identified rarely if ever at other laboratories and historically neither before nor since at the laboratories in question. In both cases, although there was a suggestion of error from the relatively random nature of the coliform detection, much detective work was required before the problems were solved. This was largely due to the intermittent and unpredictable nature of the growth.

The first organism was *Klebsiella ozaenae* which was eventually proved to be growing in the glue of the bottle label. The type of label in use at the time needed to be wetted (in a shallow tray) before sticking on the bottle. If overwetted the glue would run down the bottle. This problem would manifest itself as a rash of failed samples over one or two days but which would then not reappear for several weeks. Although the growth could potentially occur every day, in practice the right combination of overwetting and storage only allowed it to occur at around one day in twenty. Bottles that had the appropriate, albeit undefined, conditions could have tens of thousands of *K.ozaenae* on the outside of the bottle. Dip samples from service reservoirs, now no longer acceptable for regulatory sampling , were particularly susceptible to contamination. Also, inevitably (although surprisingly infrequently), contamination would occur in the laboratory during the membrane filtration process. The origin of the *K.ozaenae* was never established. It was never isolated from unwetted labels although it may have been an occasional contaminant and it was never isolated from the skin of laboratory personnel, although this again does not necessarily rule out its presence.

The second unusual species was *Escherichia adecarboxylata* (now *Leclercia*). This organism was growing as a biofilm inside the manifold of the membrane filtration apparatus in the laboratory. Although the growth was actually on the suction side of the membrane filters, the valve seals were worn and small amounts of water containing tens of thousands per ml could seep out onto the bench. Under these circumstances some contamination was inevitable, although as in the previous study 95% or more of samples were handled without any contamination.

Situations in which contamination of samples during sampling have also been identified. In one case coliform growth was found in plastic purpose built sampling crates and in another from growth in a plastic bucket used by a sampler to carry his sampling bottles and accoutrements. Unusual organisms were not involved here but in both cases there was a repeated combination of more commonly isolated organisms, *Klebsiella pneumonia* and *Enterobacter agglomerans*, found at unrelated sample points over a short period. The same combination of biochemically identical strains (same API identification number) was then tracked to the offending source.

6 LABORATORY CONFIRMATION OF COLIFORMS

The standard UK water industry requirement for confirmation of coliforms (7) requires subculture of coliforms isolated from treated water at either 37^0 or 44^0 C to agar then to lactose peptone water and tryptone water. Production of acid from lactose and indole from tryptophan is regarded as sufficient identification of *E.coli*, the designated faecal coliform. Further identification is considered optional. However, if the further identification is done routinely, using API 20E or other method, then the confirmation techniques become unnecessary. This can not only save a significant amount of work, but provides a more accurate identification; identifying *E.coli* that do not grow at 44^0 (10-15% of strains) and eliminating *Klebsiella* strains that produce indole from tryptophan - 15% of strains (6).

References

1. Drinking Water 1994. A Report by the Chief Inspector Drinking Water Inspectorate. HMSO, London, 1995.
2. United Kingdom Water Industry Research Limited, 1 Queen Anne's Gate, London. Microbiological growth in domestic pipework and fittings 1995. Report No DW004/B.
3. United States Environmental Protection Agency, Washington DC 20460. Control of biofilm growth in drinking water distribution systems 1992. Report 625\R-92\001.

4. United Kingdom Water Industry Research Limited, 1 Queen Anne's Gate, London. Health significance of bacteria growing in water distribution systems 1994. Report No DW02/A.

5. Holmes B, Gross RJ. Coliform bacteria; various other members of the Enterobacteriaceae 1990. In: Parker MT, Duerden,BI, editors. Topley and Wilson,s Principles of Bacteriology, Virology and Immunity, 8th ed, volume 2. Edward Arnold, London, 415-441.

6. American Water Works Association Research Foundation, Denver, CO 80235. Comparison of the Colilert Method and Standard Fecal Coliform Methods 1994. Report No 1P-5C-90647-5/94-CM.

7. Report on Public Health and Medical Subjects No 71. Methods for the examination of Waters and Associated Materials. HMSO, London, 1994.

Chapter 28
Occurrence Of STh Toxin Gene In Wastewater

Robin K. Oshiro and Betty H. Olson

Environmental Microbiology and Genetics Laboratory
Department of Environmental Analysis and Design
University of California, Irvine, California, USA

1 INTRODUCTION

Pathogenic *Escherichia coli* species capable of producing diarrhea can be transmitted via water through the fecal-oral route of infection. Diarrhetic *E. coli* are known to carry a variety of virulence factors such as adherence factors and fimbriae. One of the most well known of these factors is toxin production. A variety of toxins are produced by *E. coli*, including hemolysins, Shiga-like toxins (SLT), cytolethal distending toxins (CDT), cytotoxic necrotizing factor toxins (CNF), and heat labile (LT) and heat stable (ST) toxins. Many of these toxins can be further divided into distinct categories based on their phenotypic characteristics and nucleotide sequences (Mühldorfer and Hacker, 1994)

In particular, the ST toxins can be divided into STI and STII type toxins based on solubility in methanol (STI is soluble) and activity in infant mouse intestine (STI is active). STI type toxins can further be divided into STIa (STp) found in a variety of animals including exotic animals (Gulati *et al.* 1992), farm animals (Nagy *et al.* 1990, Shin *et al.* 1994, Wasteson *et al.* 1988, Woodward and Wray 1990) and humans (Gyles 1992) and STIb (STh) found exclusively in humans. Based on previous research conducted in our laboratory (Martins *et al.* 1992), it was thought that the incidence of STh could have potential as an indicator of human versus animal fecal sources.

A wide variety of animal fecal samples (251) and strains (123) of exotic, farm and household pet fecal sources from national as well as international sources were tested for the STh toxin gene (manuscript in preparation). The STh toxin gene was not detected in these samples either by DNA extraction followed by PCR amplification (fecal samples) or by hybridization with a labelled probe (*E. coli* stains isolated from fecal samples).

This paper discusses sewage samples that were assayed for this toxin gene using PCR. The purpose of this study was to determine whether the STh toxin gene could possibly be used to differentiate between human and animal *E. coli*.

2 MATERIAL AND METHODS

2.1 Sample collection

Unchlorinated sewage samples were collected from various sewage treatment plants in Southern California in 1 gallon clean polypropylene containers and transported on ice to the laboratory and were processed within eight hours of reaching the laboratory. Sewage types included raw influent (either prior to or following microscreening) and final effluent.

2.2 E. coli enumeration

Sewage samples were enumerated for *E. coli* by filtering aliquots of PBS diluted sample onto a membrane (Gelman GN-6) which was then placed on mTEC media (Difco) and incubated at 35°C for 2 hours, followed by incubation for 20 ± 2 hours at 44.5°C in a water bath. The membrane was transferred to a urease reagent saturated pad and yellow (urease negative) colonies were counted as *E. coli* colonies.

2.3 DNA extraction

DNA from raw sewage pellets or sewage having visible particulate matter (1 liter sewage spun for 90 minutes at 13,500 rpm) was extracted using one of the following two methods involving either phenol-chloroform or glass bead extraction. For phenol-chloroform extraction, the pellet resulting from the 1 liter sample was removed from the centrifuge bottles and the remaining parts of the pellet were triturated off with sterile TE (pH 8.0). The pellet and wash buffer was placed in a 2 ml centrifuge tube, with 0.3 g zirconium beads, 700 μL TE (pH 8.0), 700 μL TE (pH 8.0) buffered phenol and 50 μL 20% SDS. The tube was bead beated (BioSpec Products, Bartlesville OK) for 50 seconds at medium speed then spun for one minute at 6000 x g. The supernate was removed to a fresh tube, phenol extracted and isopropanol precipitated, and the resultant pellet was resuspended in sterile distilled water and frozen at -70°C until used for PCR amplification.

DNA extraction was also done on the above sewage using a modification of a glass bead method (Stacy-Phipps *et al.* 1995). The sewage was centrifuged as above, the pellet was placed in a 2 mL slick centrifuge tube (PGC Scientific, Gaithersburg MD) and spun for 1 minute at 2,800 x g. The supernate was removed to 2 mL centrifuge tubes and spun at 14,000 x g for 5 minutes. The resultant pellets for each sewage sample were pooled by resuspension in 1 mL guanidine thiosulfate lysis buffer (5.3 M guanidine thiosulfate, 10 mM DTT, 1% Tween 20, 0.3 M sodium acetate, 50 mM sodium citrate (pH 7.0)). 7.6 μL 0.5 M NaCl and 53 μL CTAB/NaCl (10% CTAB in 0.7 M NaCl) was added prior to incubation for 10 minutes at 65°C. 100 μL glass beads (National Scientific Supply, San Rafael CA) was added and the mix was incubated at room temperature with constant shaking for 15 minutes followed by centrifugation at 14,000 x g for 1 min. The pellet was resuspended in a solution containing 50% EtOH, 10 mM Tris HCl (pH 7.5), 100 mM NaCl and spun as above. These two steps were repeated for a total of three washes. The pellet was dried for 15 min at room temperature, resuspended in 100 μL sterile distilled water, incubated at 50°C for 5 minutes with occasional shaking and spun at 16,000 x g for 2 minutes. The eluate was transferred to a fresh tube and frozen at -70°C until used for PCR amplification.

Final effluent samples lacking particulate matter were filtered through an HVLP membrane (Millipore, Bedford MA). The membrane was placed in a sterile tube and vortexed in the presence of 1 mL guanidine thiosulfate lysis buffer as in the glass bead method above for 1 minute. The guanidine buffer was removed and further processed as in the above glass bead method.

2.4 PCR amplification

The DNA (50 ng) was PCR amplified (Perkin Elmer, Norwalk CT) in a master mix that included 10 mM Tris HCl (pH 8.3), 50 mM KCl, 2.5 mM $MgCl_2$, 0.2 mM dNTP's, 10 pmole of each primer and 2.5U *Taq* DNA polymerase. A single one minute cycle at 94°C was followed by 35 cycles of 30 sec at 94°C, 30 sec at 43°C and 30 sec at 72°C and then by a single cycle of 1 min at 72°C. The 167 bp amplicon was visualized on a 2% agarose gel and verified by cutting with Hinf I restriction enzyme and (24, 26 and 117 bp bands).

2.5 Recovery Efficiency testing

Samples from treatment plants in Southern California were collected and assayed a number of times. For these samples, efficiency tests were conducted to determine the number of STh toxin genes carrying *E. coli* that were necessary for a positive PCR amplification. The sewage was disinfected by treatment with sodium hypochlorite (20 mg/L), and subsequently dechlorinated with sodium thiosulfate (0.02 % final concentration) and then inoculated with dilutions of a known STh positive *E. coli* strain. The samples were spun and further processed using the glass bead method of DNA extraction.

3 RESULTS AND DISCUSSION

In order to determine if the STh fragment was unique a number of animal fecal samples were screened to determine STh occurrence. The results obtained to date are all negative. The types of animals screened and the number of samples are shown in Table 28.1.

Once the STh fragment proved negative for animal samples. We needed to find out its occurrence in sewage samples. All sewage treatment plants used in this research received primary and secondary treatment. The secondary treatment was usually activated sludge, but in at least one plant trickling filters were employed as the secondary treatment phase.

Raw sewage samples were found to have an approximate mean *E. coli* content of 5 x 10^5 CFU/mL. At the beginning of this study, DNA was extracted from 100 mL samples and PCR amplified for the STh gene. None of the seven samples thus tested were found to be

Table 28.1 *Animal types screened and number of isolated colonies originating from animal sources and animal fecal samples tested.*

Animal type	# Isolated colonies originated from animal sources	# Animal fecal samples
exotic zoo	13	54
farm – bird	18	2
cow	34	32
goat	8	2
horse	8	6
pig	14	5
sheep	8	5
steer	2	5
household pets	18	13

Table 28.2 *Description of Sewage Treatment Plant Facilities and STh Occurrence in Raw Influent and Treated Effluent*

Location of Sewage Treatment Facilities by State	Number of service connections or population served	MGD Received	Percentage of domestic sewage	Number of samples	% positive for STh gene (1 liter sample size)
California raw influent final effluent	11,700-48,152	2.5-12	80-97	62 33	100 100
Kentucky raw influent final effluent	50,000	NA	85	1 1	100 100
Ohio raw influent final effluent	NA	NA	NA - 90	2 2	100 100
Michigan raw influent final effluent	5,000 -95,000 people served	2.2-32.5	35-70	2 2	100 100

positive for this gene, and as a result the sample size was increased to 1 liter. The results for the one liter sample size are shown in Table 28.2.

Recovery tests were conducted for those sites where multiple samples were collected. The Irvine Ranch Water District is an example of one such location. This district serves approximately 100,000 people through 48,152 service connections. The sewage is composed of 95% domestic waste and approximately 5% commercial waste. The raw sewage receives primary and secondary treatment. Water to be used for reclamation purposes receives filtration and chlorination. The minimum recovery was 300 CFU for IRWD raw influent. Therefore a minimum of 0.00006% of the total *E. coli* community carry this toxin gene in IRWD raw influent per liter, discounting viable but nondetectable *E. coli* and extraneous DNA in the sewage.

Initial results using 100 mL produced negative results so the sample size was increased to 1000 mL. Once the larger sample size was used all results from raw and secondary treated wastes resulted in positive results. In order to determine the minimum sample size needed seeded tests were conducted using volumes of 1,000, 500 and 250 ml. The same sample was split into 4 equal parts (250 mL each), 2 equal parts (500 mL) each and 1 of 1,000 mL. These seeded samples were processed in the same manner as described above. Three of the four 250 mL samples were positive, both of the 500 mL samples and the 1,000 mL sample were positive for STh. This finding suggests that the sample size could be reduced to 500 mL, but also suggests that the same assay should be carried out in difference regions, because the number of *E. coli* entering the waste stream may differ geographically. This finding also suggests that an most probable number approach could be developed to provide quantitation of the occurrence of these *E. coli*. This information will be used to create a simple binomial model of predictability of occurrence.

References

Gulati, B.R., V.K. Sharma and A.K. Taku. 1992. Occurrence and Enterotoxigenicity of F17 Fimbriae Bearing *Escherichia coli* From Calf Diarrhoea. Vet. Record 131:348-9.

Gyles, C.L. 1992. *Escherichia coli* Cytotoxins and Enterotoxins. Can. J. Microbiol. 38:734-46.

Martins, M.T., I.G. Rivera, D.L. Clark and B.H. Olson. 1992. Detection of Virulence Factors in Culturable *Escherichia coli* Isolates From Water Samples by DNA Probes and Recovery of Toxin-Bearing Strains in Minimal *o*-Nitrophenol-ß-D-Galactopyranoside-4-Methylumbelliferyl-ß-D-Glucuronide Media. Appl. Environ. Microbiol. **58**:3095-3100.

Mühldorfer, I. and J. Hacker. 1994. Genetic Aspects of *Escherichia coli* Virulence. Microb. Pathogen. 16:171-81.

Nagy, B., T.A. Casey and H.W. Moon. 1990. Phenotype and Genotype of *Escherichia coli* Isolated From Pigs With Postweaning Diarrhea in Hungary. J. Clin. Microbiol. 28:651-3.

Shin, S.J., Y.-F. Chang, M. Timour, T.-L. Lauderdale and D.H. Lein. 1994. Hybridization of Clinical *Escherichia coli* Isolates From Calves and Piglets in New York State With Gene Probes for Enterotoxins (STaP, STb, LT), Shiga-like Toxins (SLT-1, SLT-II) and Adhesion Factors (K88, K99, F41, 987P). Vet. Microbiol. 38:217-25.

Stacy-Phipps, S., J.J. Mecca and J.B. Weiss. 1995. Multiplex PCR Assay and Simple Preparation Method for Stool Specimens Detect Enterotoxigenic *Escherichia coli* DNA During Course of Infection. J. Clin. Microbiol. 33:1054-1059.

Wasteson, Y., Ø. Olsvik, E. Skancke, C.A. Bopp and K. Fossum. 1988. Heat-Stable-Enterotoxin-Producing *Escherichia coli* Strains Isolated from Dogs. J. Clin. Microbiol. 26:2564-6.

Woodward, M.J. and C. Wray. 1990. Nine DNA Probes for Detection of Toxin and Adhesin Genes in *Escherichia coli* Isolated From Diarrhoeal Disease in Animals. Vet. Microbiol. 25:55-65.

Chapter 29
The Influence Of Sodium Hypochlorite Concentration On Retention Of *E. Coli* In A Gram Positive Biofilm

B. Daly[1], W.B. Betts[1] & J.G. O'Neill[2]

[1] Department of Biology, University of York, PO Box 373, York, YO1 5YW, U.K.
[2] Yorkshire Water, Water Quality, PO Box 201, Bradford, BD1 5PZ, U.K.

1 INTRODUCTION

Populations of micro-organisms can become established on almost any surface exposed to waters populated with microbes (Whitham & Gilbert, 1993). The immobilised cells grow, reproduce and produce extracellular polymeric substance (EPS) that frequently extends from the cell forming a mass of tangled fibres lending structure to the entire assemblage, which is called a biofilm (Turakhia & Characklis, 1989).

Several studies have been carried out on the abilities of biofilms to resist traditional bactericidal methods, such as chlorination (Douglas & vanNoort, 1993). Substantially reduced sensitivities to chlorination have been shown for *Escherichia coli*, *Salmonella typhimurium*, *Yersinia entercolitica*, *Shigella sonnei and Klebsiella pneumoniae* absorbed to carbon granules (LeChevallier *et al.*, 1988).

The major hypotheses to explain this resistance are (Brown & Gilbert, 1993):

- The glycocalyx or EPS excludes and/or influences the access of antimicrobial agents to the underlying organisms,

- For chemically reactive antimicrobial agents such as the halogens and sulphydryl-interactive agents (e.g. isothiazolones), and for physically adsorbed agents in low concentrations, the surface regions of the glycocalyx and outlying cells react with and quench the biocides,

- Limited availability of nutrients in the biofilm forces a slowing of growth and adoption of atypical phenotypes thus rendering the antimicrobial agents less effective,

- Attachment causes derepression of genes associated with a sessile existence which coincidentally affect antimicrobial susceptibility.

This paper describes preliminary work using a small scale biofilm generator to grow a Gram-positive biofilm subsequently seeded with the faecal indicator bacterium *E. coli*. The research examined the ability of biofilm component organisms to survive and reproduce when subjected to increasing levels of free chlorine. This type of investigation should increase the knowledge of potential problems caused by faecal pathogens being incorporated into biofilms in water pipes and assist in understanding the mechanisms underlying the "protection" of pathogens afforded by biofilms.

2 METHODS

2.1 Biofilm development

A biofilm generator system (Harrison *et al.*, 1995) (Figure 29.1) was sterilised by flushing with disinfectant (Virkon, Antec Int.) for 24 h, followed by sterile distilled water. Nutrient broth (1/8 strength) was introduced into the system and allowed to replace the water. 5 ml volumes of overnight cultures of 2 Gram-positive bacterial strains isolated from tap water were injected into the system at point A. This inoculum was circulated for 1 h before the waste pump (Roussel) was switched on. After a 24 h period of incubation 5 ml of an *E. coli* 11459 (NCIMB) suspension (approx 8 x 10³ cfu) were inoculated into the system. Nutrient broth (1/8 strength) was continuously circulated (Watson-Marlow pump) with the waste pump removing spent medium at a rate of 150 ml h⁻¹. The biofilm was allowed to develop for 5 d.

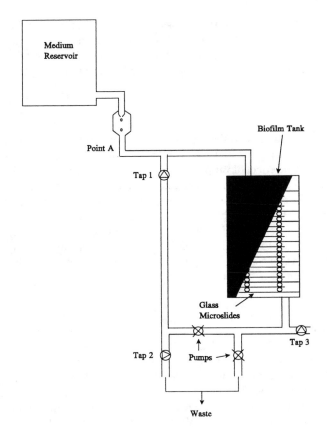

Figure 29.1 *Biofilm generator system*

2.2 Sodium hypochlorite treatment

Sodium hypochlorite (12% solution from GPR) was aseptically added to the sterile nutrient broth reservoir to give the required free chlorine concentration, which was measured using a chlorine probe (Lovibond, 2000 comparator MK.ll). To ensure >85% dissociation of the free chlorine as hypochlorous acid the pH of the reservoir was monitored to ensure it remained in the 7-7.4 range.

2.3 Sampling of *E. coli*

Tap 1 was closed and tap 2 opened for 10 min to allow fresh medium to flush through the system. This removed planktonic *E. coli* and allowed analysis of biofilm-derived *E. coli*. Samples were taken from tap 3 every 30 min for the first 4 h and hourly thereafter. These were decimally diluted in sterile water and plated onto MacConkey agar. The plates were counted after incubation at 37°C for 24 h. After addition of the sodium hypochlorite, 15 h contact time was allowed before sampling took place.

2.4 Effect of sodium hypochlorite on planktonic *E. coli*:

E. coli cultures were initially grown in flasks containing 100 ml nutrient broth at 37°C for 24 h. Diluted samples were plated on MacConkey agar and incubated. After sampling, sodium hypochlorite was added to the flasks to give final free chlorine concentrations of 0.225, 0.25 and 0.275 mg l^{-1}. These were incubated at 37°C for 15 h and were sampled again.

3 RESULTS

E. coli derived from biofilm showed significant differences in survival relative to cells grown in the planktonic phase when exposed to sodium hypochlorite.

Figure 29.2 shows the effect of various concentrations of free chlorine on flask cultures of *E. coli* grown in planktonic mode. A free chlorine concentration of 0.25 mg l^{-1} was sufficient to kill an overnight culture of *E. coli*, whilst a lower free chlorine concentration of 0.225 mg l^{-1} resulted in dramatic inhibition of the culture.

Figure 29.3 presents counts of *E. coli* derived from biofilm subjected to different free chlorine concentrations up to a maximum of 1.0 mg l^{-1} (i.e. four times the concentration required to kill the planktonic culture using the same contact time of 15 h). This shows the survival of large numbers of *E. coli* in the biofilm generator system at this high free chlorine concentration.

4 DISCUSSION

Biofilm derived *E. coli* was capable of survival in large numbers at free chlorine concentrations several times higher than that required to kill a planktonic culture. The 10 min flush with fresh medium almost certainly ensured that the majority of *E. coli* originated from the biofilm. These *E. coli* must have been protected in some way, either by a component of the biofilm or by their own physiological characteristics. Various proposed methods of biofilm resistance were described above and it is conceivable that any of these could result in sloughed off bacteria retaining their resistance to free chlorine during the sampling procedure.

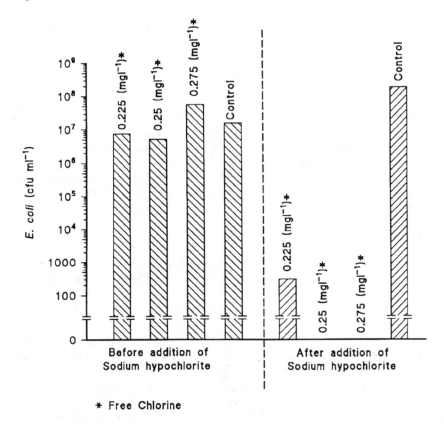

Figure 29.2 *Survival of planktonic E. coli after exposure to sodium hypochlorite*

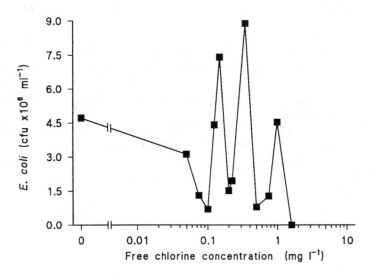

Figure 29.3 *Recovery of E. coli from a biofilm model treated with sodium hypochlorite*

These results are important in the context of the problems caused by biofilms in water distribution networks. *E. coli is used to indicate the presence of faecal material of human or animal origin (Anon., 1994)* and strains of the organism can be pathogenic. While conventional water treatment methods are directed towards elimination of planktonic microbes it is apparent from these results that if a biofilm does form it presents the opportunity for small numbers of pathogens to adhere and reproduce in large numbers despite relatively high free chlorine concentrations. It has also been shown that the protozoan parasite, *Cryptosporidium parvum,* which is not entirely eliminated by conventional water treatment (including a chlorination stage), can exist attached to surfaces in a model biofilm system (Rogers & Keevil, 1995). Sloughing from biofilms might release aggregations of cells containing pathogens into the planktonic phase of a potable water distribution system and subsequent consumption could lead to infection of consumers. The work in this paper further indicates the potential public health risks presented by biofilm formation in water distribution systems.

Acknowledgements

The authors wish to thank Yorkshire Water Services Ltd. for postgraduate studentship support and Adrian Harrison & Andrew Brown for assistance with the biofilm generator system.

References

ANONYMOUS (1994) The Microbiology of Drinking Water, Part 1 - Drinking Water. Report on Public Health and Medical Subjects No. 71, Methods for the Examination of Waters and Associated Materials. London: HMSO.

BROWN, M.R.W. & GILBERT, P. (1993) Sensitivity of biofilms to antimicrobial agents. *Journal of Applied bacteriology Symposium Supplement.* **74**, 87S-97S.

DOUGLAS, C.W.I. & VAN NOORT, R. (1993) Control of bacteria in dental water supplies. *British Dental Journal.* **174**, 167-174.

HARRISON, A.B., BROWN, A.P., BETTS, W.B. & O'NEILL, J.G. (1995) A laboratory biofilm generator adaptable for recirculatory or flow-through operation. *Microbios*, **84**, 53-62.

LeCHAVALLIER, M.W., COUTHON, C.D. & LEE, R.G. (1988) Inactivation of biofilm bacteria. *Applied & Environmental Microbiology*, **54**, 2492-2499.

ROGERS, J. & KEEVIL, C.W. (1995) Survival of *Cryptosporidium parvum* oocysts in biofilm and planktonic samples in a model biofilm. In *Protozoan Parasites and Water.* eds. Betts. W.B., Casemore, D.P. Fricker, C.R. Smith, H.V., Watkins, J., pp 143-145. Cambridge: Royal Society of Chemistry.

TURAKHIA, M.H. & CHARACKLIS, W.G. (1989) Activity of *Pseudomonas aeruginosa* in biofilms: Effect of calcium. *Biotechnology & Bioengineering.* **33**, 406-414.

WHITHAM, T.S. & GILBERT, P.D. (1993) Evaluation of a model biofilm for the ranking of biocide performance against sulphate-reducing bacteria. *Journal of Applied Bacteriology.* **75**, 529-535.

Subject Index